I0049593

GAUSS NODES RE

Numerical Integration theory radically simplified and generalised.

Author: Rob Porter

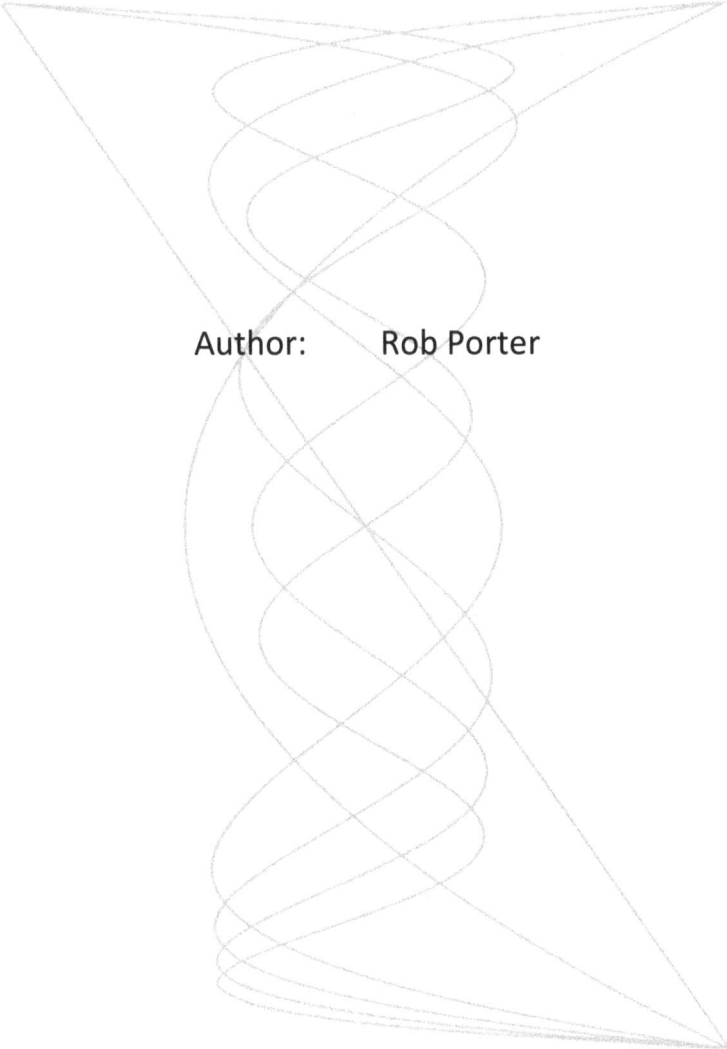

GAUSS NODES REVOLUTION

Copyright © Rob Porter

First published: 2023

ISBN: 978-0-6456776-5-2 paperback

ISBN: 978-0-6456776-6-9 e-book

Copyright Notice: All rights reserved. Without limiting the rights under copyright reserved above, no part of this publication may be reproduced, stored in or introduced into a database and retrieval system or transmitted in any form or by any means (electronic, mechanical, photocopying, recording or otherwise) without the prior written permission of the owner of the copyright.

Disclaimer: The data and information given in this document is not guaranteed to be accurate. Users should verify the data themselves before trusting the data. The author assumes no responsibility or liability whatsoever.

CONTENTS

iv

PREFACE

How can my claim of a Gauss Node Revolution can be true given that Gauss nodes have been around for so long?

Although it is true that Gauss nodes were invented a long time ago, who (other than Gauss and a few other enthusiasts) would try to find Gauss nodes without the help of the modern computer. And so it is that nothing would have been possible without Microsoft Excel (with VBA and the 'xlPrecision' add-in).

Wolfram Alpha too has been a help as has R. W. Hamming's book "Numerical Methods for Scientists and Engineers'(Dover Publications, 2nd Edition) and Wikipedia.

GAUSS NODES REVOLUTION

1. INTRODUCTION

Gauss-nodes are special sample points (x-values) and associated weights devised to find an approximation of the area under a smooth y =f(x) curve, typically between x=-1 and x=1. The idea is that multiplying each of the y-values at the sample points, by the corresponding weight-value and adding the results will give a reasonable estimate of the area under the curve. The process of using data to estimate the area under a curve is sometimes called 'numerical integration', 'integration' or 'quadrature'.

The mathematics behind Gauss nodes can seem too complex to be interesting. According to the first paragraph of Wikipedia: *Gaussian quadrature*: "In numerical analysis, a quadrature rule is an approximation of the definite integral of a function,... An n-point Gaussian quadrature rule, named after Carl Friedrich Gauss, is a quadrature rule constructed to yield an exact result for polynomials of degree 2n – 1 or less by a suitable choice of the nodes x_i and weights w_i for i = 1, ..., n. Modern formulation using orthogonal polynomials was developed by Carl Gustav Jacobi 1826.[2]... It can be shown (see Press, et al., or Stoer and Bulirsch) that the quadrature nodes x_i are the roots of a polynomial belonging to a class of orthogonal polynomials (the class orthogonal with respect to a weighted inner-product). This is a key observation for computing Gauss quadrature nodes and weights."

My alternative 'vector' explanation of Gauss-nodes is more geometrical than the traditional mathematical approach. The vector approach does not touch on the concept of orthogonal polynomials for example. I use the term *vector* or *vector function* to mean the simple shapes like 1, x, x^2, x^3 ... e^x, $e^{(2x)}$ etc, and some times more complex shapes. Traditional Gauss-nodes are based on polynomial vectors.

A vector function has an external linear coefficient, a variable, and an internal shape parameter, though it may seem to not have an external linear coefficient or an internal shape parameter when they have been set equal to 1. We can think of the y=1 vector as an exception because it does not even have a variable, but we can fix that by writing it as $1^{\wedge x}$ or $x^{\wedge 0}$.

If we use 'b' to represent the internal shape parameter and 'a' to represent the external linear coefficient and 'x' to represent the variable then we can use a*V(x, b) to represent some vector. In theory then we can construct a linear combination of such vectors the form: a_1*V(x, b_1) + a_2*V(x, b_2) +...to map some target function t(x) at various sample points.

Polynomial vectors are different because the shape parameter b (the polynomial index) can only take on positive integer values and zero. Polynomial vectors are a special kind of $x^{\wedge b}$ power vector. When the linear coefficient is set to one, all polynomial vectors except for the y=1 vector pass through the points (0,0) and (1,1). All polynomial vectors except for the y=1 and the y =x vectors are tangent to the x-axis at the origin whereas all power vectors with an index greater than 0 and less than 1 are tangent to the y-axis at the origin (for x >= 0).

Newton and Lagrange interpolation methods allow us to determine the linear coefficients to apply to the polynomial vectors without the need to solve simultaneous equations. And we do not need to find the shape parameter values because they are predetermined to follow the polynomial rule which requires that we use the smallest positive integer indices first, starting from b=0. This means that a polynomial mapping of some data, unlike say an exponential mapping found using difference equations, is a single parameter (linear coefficient only) mapping. Instead of finding special values of the shape parameters b_1, b_2, b_3, ... that map the data as we would for a two-parameter vectors we adopt the values given to us by the polynomial rule and find only the linear coefficients. And finding the linear coefficients is a simple exercise since we can use Newton or Lagrange interpolation methods, or we can solve the linear simultaneous equations.

After we have found the linear coefficients required to map our data, we need only sum the integrals of each vector to get an approximation for the integral of the curve.

The concept of 'weights' used with Gauss-nodes and 'integration factors' used with Newton-Cotes rules means that we can skip the step of finding the linear coefficients required to map the shape of the curve. The use of 'weights' and 'integration factors' means we need only multiply each sample point y-value by the corresponding weight or integration factor, and then sum the result to get an approximation of the area under the curve.

Although traditional Gauss-nodes are based on polynomial vectors, there are traditional Gauss-nodes that are adapted to integrating target curves that are better represented by the product of some simple non-polynomial weight-function and a polynomial, say: $f(x) \sim WT(x)*\{a_0+a_1 x+ a_2 x^2+a_3 x^3+... \}$. Gauss–Laguerre and Gauss-Hermite nodes use the weight-functions e^{-x} and e^{-x^2} in conjunction with polynomial vectors. The inclusion of weight-functions means we can expand the scope of the target functions to include functions that are not easily mapped with polynomials.

Gauss-Legendre nodes, the most common type of Gauss nodes (which I refer to as 'CGNs' and 'conventional Gauss nodes') do not have a weight-function ($WT(x)$ is set to 1).

Why do we need Gauss-node sample points given that we can easily find weights that are appropriate to any set of sample points? Why not use equally spaced points for example? The issue with evenly spaced nodes is that even when the target curves are simple curves, polynomials do not always converge towards the target curves at locations in-between the nodes. For example, if we map the Runge curve $1/(1+ 25\ x^2)$ with equally spaced nodes between x=-1 and x=1; then as we add more equally spaced nodes the in-between errors increase—see: Wikipedia: Runge Phenomenon.

Newton-Cotes (NC) rules are integration rules based on sample points one unit apart. The Runge Phenomenon means we cannot use high-order NC nodes to integrate curves like the Runge curve. Gauss-nodes help to address this problem.

2. GNR ADVANTAGES

Benefits of the Gauss Nodes Revolution (vector) approach to Gauss nodes include:

1. INSIGHT: The 'vector approach' to Gauss-nodes provides a better understanding of what Gauss-nodes are and why they work.

2. MORE GENERAL: The vector approach allows the adoption of alternative non-polynomial vectors. For example, we could use the Gauss–Laguerre weight-function *as the vector* and some other function as the weight-function.

3. SIMPLICITY: Given that polynomials are the simplest of all vectors to work with, it may seem a little daunting to consider non-polynomial vectors as our Gauss-node vectors. However, the vector approach using non-polynomial or polynomial vectors is easier to follow than the traditional approach to Gauss-nodes.

4. METHODS FOR FINDING GAUSS-NODES: In addition to understanding how and why Gauss-nodes work, some readers will want to find their own Gauss nodes (to suit a special set of target curves). Although a multi-variable iteration process can often find any nodes we want, wherever possible I will describe a one-step non-iterative mathematical method.

Disclaimers & Clarifications

My reference to a 'one-step non-iterative mathematical method' to find Gauss-nodes, means only that we can replace the *multi-variable* iteration process with a mathematical process. Finding a set of Gauss nodes will require many steps and there are not that many instances when we can avoid the need for single-variable iteration, say to find the roots of equations. 'Multi-variable iteration' means we have more than one independent variable acting on one or more dependent variables and we hope to find the preferred values of all the independent variables by taking numerous small steps in the right direction.

Some of the mathematical methods will require familiarity with differentiation, integration, Taylor expansions and the simplest kinds of DE's (differential equations and difference equations with solutions in the form of linear combinations of exponential terms so that we know the solution structure in advance, and we need only find the coefficients.)

For those amongst us who believe that more than one simultaneous equation is too many, matrix inversion will be required. There have been occasions when I have found that Microsoft Excel will not give sensible answers for inversion of even a 16x16 matrix. My solution has been to use the xlPrecision add-in.

I do not claim to answer the question of 'best Gauss-nodes'. If we want to integrate low order polynomials, then we need CGN nodes. Of course, if we were truly only interested in integrating low order polynomials then we could use NC nodes of a sufficiently high order.

Although I will not be claiming a 'best Gauss-node' solution, I will explore one or two error-criteria and of course there is no way to avoid the us of a *multi-variable* search algorithm if we want to find Gauss-nodes that satisfy some error-criterion.

3. UNDERSTANDING CGN's

A Multi-variable Search Algorithm

Thinking about how to design a multi-variable search algorithm to find CGN's will help our understanding of CGN's (Conventional Gauss-Legendre nodes).

Suppose we commence with five x-nodes located at say $\{0, \pm1/2, \pm3/4\}$ with the $x_0=0$ node being a fixed-node (or predetermined node) but the other x-nodes being free nodes and we want to move the free nodes around to change the results. One obvious but very important fact is that we only have 2 free x-nodes x_1 and x_2, because (apart from a change in sign) symmetry dictates that the LHS node values will determined by RHS node values. The same is true of the weights except for the centre node weight which is a free parameter. So, we initially have $x_1= 1/2$ and $x_2 =3/4$ as our 2 free x-nodes and we have 3 free weights, so we have a total of 5 free parameters.

Let's call vectors x^2 & x^4 the 'primary-vectors'. Using matrix inversion to solve the linear simultaneous equations, we can easily find the 2 linear coefficients (i.e., the weights) associated with the initial x-node locations that will integrate the primary-vectors exactly.

Because we have a fixed centre node, the x^0 ($y=1$) vector cannot be counted as a primary vector. Because we have a fixed centre node, and the x^0 vector is the only vector that has a non-zero value at $x=0$, the centre-node weight must be associated with the x^0 vector and determined last. Essentially, the centre node weight will be the value that makes all the weights sum to 1 for the RHS nodes (or 2 for the combined RHS & LHS nodes) as this is the condition required to ensure that the nodes integrate the $y=1$ line exactly.

We now identify a set of secondary-Gauss-vectors, say $x^{\wedge 6}$ and $x^{\wedge 8}$ (one for each free node) and estimate the area for each of the secondary-Gauss-vectors using the weights derived from the primary-Gauss-vectors. We make the area-estimate errors for the secondary vectors, the dependent variables (y_1 and y_2) and the two free x-node values x_1 and x_2, the independent variables. If we change the independent variables to {0, ±0.499, ±0.749} say, then our matrix inversion will give us new weights that will integrate the primary-vectors exactly and new area-estimate errors for our secondary-vectors.

The trick is now to adopt an algorithm to increment the independent variables in the direction required to reduce the secondary-vector integration errors and hope that enough small steps will get us to where we want.

In essence then, because there are an infinite number of nodes and weights that will integrate our primary-vectors exactly, we hope to be able to find the x-node sample points that will allow us to minimise secondary-vector errors. And as it turns out we can find x-nodes that will reduce the secondary-vector errors to zero. Such x-nodes and weights will be CGN Gauss nodes. For our 5 free-parameter Gauss node problem described above, the x-nodes and weights solution will be what I refer to as CGN5 nodes.

Node Symmetry

For every free x-node we have one primary vector and one secondary vector that we can integrate exactly which means 2 vectors per free x-node. The centre-node is not free, but the centre-node weight is free, and as explained above it must be associated with the y=1 vector.

There is no permanent distinction between the primary vectors and the secondary vectors; we could have made $x^{\wedge 2}$ and $x^{\wedge 4}$ the secondary Gauss vectors. However, it is important (and trivial) to note that all the primary and secondary vectors are even functions because we have skipped over the odd vectors x, $x^{\wedge 3}$ and $x^{\wedge 5}$. Providing the x-nodes and weights are symmetrical about the x-axis (i.e., that the x-values of the LHS nodes are equal to the -1 times the RHS node values and the RHS weights are identical to the LHS weights) then our nodes are guaranteed to integrate not just low order odd polynomials but all odd functions and not just approximately but exactly even when our weights and nodes include radically wrong (but appropriately symmetric) values. So then, because the integration of odd functions is so well and truly done and dusted with the adoption of symmetric nodes, why not find the RHS nodes and weights and add the LHS nodes and weights after the RHS values have been found?

Apart from the adjustment we will need to make because the RHS nodes only integrate from 0 to 1 (rather than from -1 to 1) we know that the RHS nodes must always give us the same result we get from the combined nodes when we are integrating even functions. And any function f(x) can be converted to an even function with the same are using the rule (f(x)+f(-x))/2. So, the RHS nodes are the primary shape-integrators, but we need the LHS nodes so that the combined nodes integrate all odd functions exactly without the need to convert them to even functions. (I like to identify the RHS nodes as the shape nodes but of course, this is mere convention.)

Finding CGN X-Nodes and Weights

Consider the N5 RHS Gauss-nodes copied from by Mike "Pomax" Kamermans, Gaussian Quadrature Weights-and-Abscissae webpage: https://pomax.github.io/bezierinfo/legendre-gauss.html:

CGN5 Gauss Nodes	
x-Node	Weight
0	0.28444444444444444
0.5384693101056831	0.4786286704993665
0.9061798459386640	0.2369268850561891

I have taken all my CGN nodes from the above webpage except that instead of identifying the above nodes as 'N5' nodes I identify them as 'CGN5' nodes. And instead of saying the 5 in the N5 means that nodes will integrate a polynomial up to order '2*N-1' (as per the Wikipedia citation given in the introduction), I will be saying that the 5 means there are 5 free Gauss parameters. Also, as my nodes and weights are given as only RHS nodes and weights, instead of the conventional weight 0.5688888888888889 I show half the conventional weight: 0.2844444444444444.

If we ignore all the odd polynomials (x, x^3, x^5, x^7) and focus on finding the RHS values of the CGN5 nodes then this means we want to find the values x_1, x_2 and w_1 and w_2 that make the following 4 equations true:

$$1: \quad w_1 x_1^2 + w_2 x_2^2 \quad = 1/3,$$

$$2: \quad w_1 x_1^4 + w_2 x_2^4 \quad = 1/5,$$

$$3: \quad w_1 x_1^6 + w_2 x_2^6 \quad = 1/7,$$

$$4: \quad w_1 x_1^8 + w_2 x_2^8 \quad = 1/9,$$

The numbers on the RHS are equal to $1/(b+1)$ where b is the index of the polynomial on the LHS and $1/(b+1)$ is the area of the x^b from x=0 to 1. We could express the above conditions as:

$w_1 x_1^{\wedge b} + w_2 x_2^{\wedge b} \sim 1/(b+1)$, exact for $b \sim \{2,4,6,8\}$.

Once we eliminate the odd polynomial conditions the problem of finding CGN's becomes the problem of finding the exponential curve fit to the $1/(b+1)$ curve: $w_1 x_1^{\wedge b} + w_2 x_2^{\wedge b} \sim 1/(b+1)$, (for $b\sim 2,4,6,8$). And because of the even spacing of the b-nodes (as dictated by the polynomial rule) the problem of finding CGN's is a routine difference equation problem. The independent variable (b) is spaced 2 units apart whereas conventional difference equation methodology assumes the independent variable is spaced 1 unit apart, but we can fix that by proceeding as if there was some new independent variable (say c) that is spaced 1 unit apart and then after we have found the x-nodes corresponding to variable 'c' we can take the square roots of the x-nodes to get x-nodes that correspond to variable 'b'.

Difference equation solutions are restricted solutions because they are suited to evenly spaced data samples but as it happens such restrictions fit nicely with the polynomial rule restrictions.

We now want to generalise our understanding of CGN vectors so that we can generate non-polynomial non-CGN Gauss nodes.

4. VECTOR SOLUTIONS

This section is about developing and extrapolating our understanding of CGN Gauss nodes to non-CGN Gauss nodes.

Two Parameter Vectors

The x-nodes and weights for CGN's are the parameter values of an exponential curve fit to the definite integral (A(b)) of the $x^{\wedge b}$ polynomial target-curve vector. But what are Gauss Nodes for non-polynomial vectors? (When I speak of nodes and weights being the curve fit parameters, I mean only the RHS x-nodes and weights. We can ignore the centre node and the LHS nodes because they are not free nodes. The weight associated with the centre node, if there is a centre node, is also a free parameter, but it needs special consideration.)

All vectors have at least one internal shape parameter. When we want to combine two or more vectors with different internal shape parameters to make complex shapes, we can use external linear coefficients.

When we use polynomials to map data, the practise is to follow the polynomial rule and use polynomials of the lowest order first; this means we are not selecting the shape parameter of the vectors to map the data and that the polynomial shape parameter is not a free parameter. Because a polynomial curve fit is a single-parameter (linear-coefficient only) curve-fit, that means when we have n number data that we want to map then we will need n number vectors with n different shape parameters and n linear coefficients.

The CGN5 exponential mapping of the 1/(b+1) area curve developed above is a two-parameter curve fit because both the shape parameters (the bases of the exponentials) and the linear coefficients have been selected to map the data. This means that when we have '2n' number data we hope to be able to find 'n' vectors ('n' free vector shape parameters) and 'n' free linear coefficients to map the data instead of '2n' predetermined vector shapes (polynomials) and '2n' free linear coefficients.

Regardless of whether we are using linear-coefficient-only vectors or two-parameter vectors the rule is that we expect to be able to map one point on the target curve for every free parameter. (Polynomials, like all vectors have an internal shape parameter but I like to refer to them as single parameter vectors because we typically only manipulate the linear coefficients when we are mapping data with a polynomial.)

When our CGN intersections with the $A(b)$ curve occur at $b=\{b_1, b_2, b_3,...b_{2n}\}$ for some $\{x_1, x_2,...x_n\}$ and $\{w_1, w_2,...w_n\}$ then we can say that: $w_1 x_1^{\wedge b} + w_2 x_2^{\wedge b} +...+ w_n x_n^{\wedge b} = 1/(b+1)$ for $b\sim\{b_1, b_2, b_3,... b_{2n}\}$ and that means we can integrate $\{x^{\wedge b1}, x^{\wedge b2}, x^{\wedge b3},... , x^{\wedge b2n}\}$ exactly since the integral of $x^{\wedge b}$ is exactly equal to $1/(b+1)$. In other words, CGN's derived from a 2-parameter exponential curve fit to the $A(b)$ $=1/(b+1)$ curve will integrate $x^{\wedge b}$ vectors exactly for those values of b (the b-nodes) where the two-parameter curve-fit intersects the $A(b)$ curve.

In effect CGN's use a two-parameter exponential vectors to map the $A(b)$ area curve to give us the x-node sample points and the weights. The b-nodes are indices of the polynomial and the points of intersection of the exponential curve-fit with the $A(b)$ curve. The x-nodes are the shape parameters and the bases of the exponential vectors. The weights are the linear coefficients of the exponential curve fit to the $A(b)$ curve. The two-parameter curve fit will be exact at the b-nodes. Before applying this CGN thinking to non-polynomial vectors, we want to consider the kind of vectors we might want.

Vector Qualities

If V(x, b) is some vector then we can poly-normalise it so that it passes through the points (0,0) and (1,1) by first subtracting the constant V(0,b) from it to get (V(x,b)-V(0,b)) and then dividing by the constant (V(1,b)-V(0,b)) to get the poly-normalised vector: (V(x,b)-V(0,b))/ (V(1,b)-V(0,b)). Poly-normalisation helps us to compare non-polynomial vectors with polynomial vectors. Providing V(x,b) has a suitably simple shape then after poly-normalising it, we may be able to divide the spectrum of vector shapes into 3 categories:

1. U-curve vectors (vectors similar in shape to high order polynomials) with a poly-normalised area significantly less than 1/3.

2. Mid-range $x^{\wedge 2}$ vectors (and a small range of similar shaped vectors adjacent the $x^{\wedge 2}$ vector) with a poly-normalised area of close to 1/3.

3. Bell-curve vectors (like an inverted poly-normalised version of the Runge curve) typically with a poly-normalised area significantly more than 1/3. (The inverted and poly-normalised version the Runge curve $1/(1+25*x^{\wedge 2})$ is $26*x^{\wedge 2}/(1+25*x^{\wedge 2})$).

Although the poly-normalisation of vectors is not required to create Gauss nodes, we probably want our vectors to be smooth curves. We may also want V(x, b) to be such that V(Abs(x), b) is a smooth curve at x=0. This is because the RHS nodes are the shape integrators developed from the vectors and apart from the adjustment we will need to make because the RHS nodes only integrate from 0 to 1 (rather than from -1 to 1) the RHS nodes must always give us the same result we get from the combined nodes when we are integrating V(Abs(x), b). In other words, there might not be much of an issue if our RHS nodes were designed to integrate x^3, x^5, ... because $(Abs(x))^3$, $(Abs(x))^5$,... are smooth curves, however $Abs(x)$ and small index power curves like $(Abs(x))^{0.5}$ are not smooth curves. With this 'parallel to the x-axis at the origin property' in mind, we can say that $tanh(x)$, $atan(x)$ and $sinh(x)$ are more like $y=x$ than $y=x^3$, and $y=x^3$ is more like $y=x^2$ than $y=x$.

Consider the vectors $3\,x^2$ and $4\,x^3$. Both vectors have an integral of 1 from $x=0$ to 1 and they intersect at $x=0.75$ which means the same weight is applicable to both vectors at $x=0.75$ and therefore $x_1 = 0.75$ must be a Gauss-node for these two vectors. Node symmetry guarantees that we can integrate odd functions, but in this case the combined LHS and RHS nodes will integrate x^2 and $Abs(x)^3$ exactly. And because $Abs(x)^3$ is a smooth curve maybe it makes some sense to include high order odd polynomials, but we might want to avoid making power curves with indices less than 1 the *focus* of our nodes. Not that there can ever be anything wrong with being able to integrate low index power curves but that we want to make the best use of the limited number of sample points.

Non-Polynomial Vectors

Suppose we have selected the vector $V(x, b)$. Following the reasoning used for CGN's, if we can map the $A(b) = \int_0^1 V(x, b)\, dx$, at $b=\{b_1, b_2, b_3,...b_{2n}\}$ for some $\{x_1, x_2,...x_n\}$ and $\{w_1, w_2,...w_n\}$ then we can say that: $w_1*V(x_1, b) + w_2*V(x_2, b) +...+ w_n *V(x_n, b) = A(b)$ for $b=(b_1, b_2, b_3,... b_{2n})$ and that will mean we can integrate $\{ V(x_1, b), V(x_2, b) ,V(x_n, b)\}$ exactly since the integral of $V(x, b)$ is exactly equal to $A(b)$. In other words, Gauss nodes derived from a two-parameter curve fit to the $A(b)$ curve will integrate $V(x, b)$ vectors exactly for those values of b (the b-nodes)where the curve-fit intersects $A(b)$.

We want our Gauss nodes not just to integrate all the b-node vectors exactly but also to integrate all linear combinations of such vectors exactly. If assuming 4 b-nodes we substitute the general linear combination expression y= a_1*V(x, b_1) +a_2*V(x, b_2) +a_3*V(x, b_3) +a_4*V(x, b_4) for some target curve (where b_1, b_2, b_3, b_4 are the b-nodes corresponding to the x-nodes and weights) into our Gauss x-nodes and weights rule, we will get:

w_1*(a_1*V(x_1, b_1) +a_2*V(x_1, b_2) +a_3*V(x_1, b_3) +a_4*V(x_1, b_4))
+w_2*(a_1*V(x_2, b_1) +a_2*V(x_2, b_2) +a_3*V(x_2, b_3) +a_4*V(x_2, b_4))

a_1*(w_1* V(x_1, b_1) +w_2* V(x_2, b_1)) + a_2*(w_1* V(x_1, b_2) + w_2*V(x_2, b_2)) + a_3*(w_1* V(x_1, b_3) + w_2*V(x_2, b_3)) + a_4*(w_1* V(x_1, b_4)) + w_2*V(x_2, b_4))

= a_1*A(b_1) + a_2*A(b_2) + a_3*A(b_3) + a_4*A(b_4)

Hence the integral of the combined vectors is the sum of the integrals of the individual vectors for all linear combinations, as required.

If the b-values of the target curve vectors correspond to the b-values of the intersections on the A(b) curve, then our result will be perfect. Given the restraints of Excel calculation accuracy and 15 significant digit nodes I count an error less than ~ 5*10^-15 as perfect performance which means we can get perfect performance even when we are targeting vectors with shape parameter values in-between the b-nodes. And because we are more often interested in integrating shapes rather than low order polynomials, we will not always know the b-values (the shape parameter values) of the target curve vectors and therefore we may be interested in the performance at b-values in-between and remote from the b-nodes:

- We might want to know the overall curve-fit of the two-parameter mapping of the A(b) curve

- We may want to put more focus on certain zones of the A(b) curve.

- It could be that we will be happy to allow that our Gauss-node curve-fit conditions map the low order derivatives at one or more points so that our curve fit generates tangents rather than intersections through the A(b) curve.

Such considerations suggest that there might be more to finding Gauss nodes then simply achieving the required number of curve-fit conditions to the A(b) curve.

Error Curves

A set of x-Nodes and weights are curve-fit parameters that we can use in conjunction with a vector to track the A(b) area curve for the vector we have selected. We can assess the performance of our Gauss nodes by comparing these two curves.

If we want our CGN5 nodes to integrate low order polynomials, then we might not want to truncate the 15 significant-digit parameter values. However, if we want our nodes to map smooth curves then we might want to know how the exponential curve fit tracks the A(b) curve over the entire spectrum, and instead of charting the CGN5 exponential: 0.4786286704993665 * 0.5384693101056831^{b} $+0.2369268850561891$ * 0.9061798459386640^{b} we could round the parameter values down to 2 significant-digits and chart $0.48*0.54^{b} +0.24*0.91^{b}$ versus $1/(b+1)$:

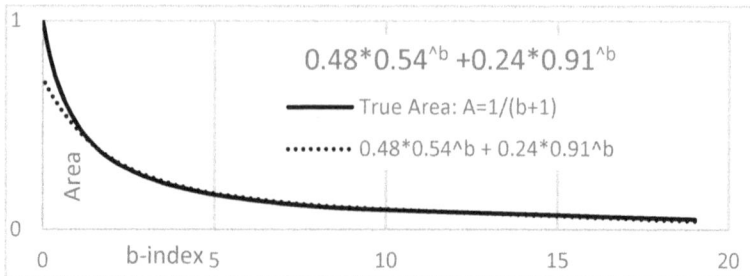

The above chart does not show the entire spectrum because it uses the b-index as the vector shape parameter to scale the units on the horizontal axis; this means the vectors with a poly-normalised area close to 1 (extreme bell-curves) are on the left and vectors with a poly-normalised area close to 0 (extreme U-curves) are towards infinity on the right.

From here on, my assessment charts will typically use a variable like the 'poly-normalised area' as the vector shape parameter on the horizontal axis so that the x-axis can be scaled from 0 to 1 with small area U-curve vectors on the left and large area inverted bell-curve vectors on the right. And from here on, instead of comparing the Gauss-Node-Area to the True-Area, the assessment charts will subtract the two curves and show a single Error Curve defined as 'Gauss-Node-Area minus the True-Area'.

If we want to see the performance of our CGN5 nodes near the 4 intersections close to b ~2,4,6,8 (or A~0.33,0.2,0.14,0.11), we need to use at least 4-digit coefficients and plot the exponential minus the target curve: $y = 0.4786*0.5385^b + 0.2369*0.9062^b - 1/(b+1)$ to get the 'Error Curve' shown below with the intersections now showing on the horizontal axis:

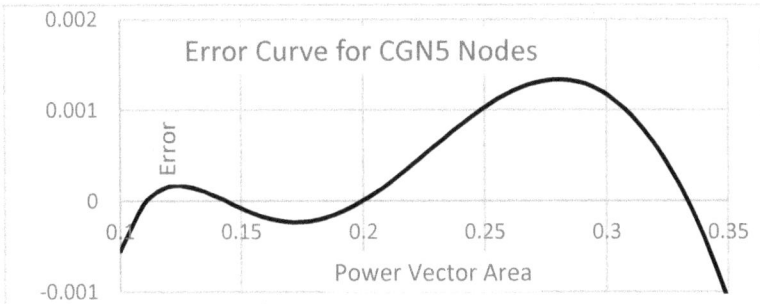

Error Curve for CGN5 Nodes

The 'Power Vector Area' on the horizontal axis identifies the power vector we are integrating with the CGN5 nodes. This means that the area value shown on the horizontal axis is effectively a vector shape parameter. For example, a power vector area of 0.2 identifies that the nodes are integrating the power vector $x^{\wedge 4}$, consequently the error is zero at 0.2. CGN5 nodes also integrate $x^{\wedge 2}$, $x^{\wedge 6}$ and $x^{\wedge 8}$ so there will also be intersections on the horizontal axis at A= 1/3, 1/7 & 1/9.

The 'estimated-area minus the true-area' error measurement means an error measurement that will remain fixed as we raise or lower the target curve relative to the x-axis; this means an error measurement that is more like a 'curve-fit error' than a 'relative-error'. (Dividing the 'curve-fit error' by the 'true-area' would give us a relative error but it would change as we raise and lower the target curve. A relative error would put more emphasis on errors at the A=0 end of the spectrum.)

The above style of Error Curve means that for all CGN's, we will see a pseudo-node at A=0 on the Error Curve, because for the A=0 extreme U-curve vector, y=0 everywhere except at x=1 and CGN's do not sample the y-value at x=1.

The above Error Chart restricts the range of b-values to, b~1.86 to 9 to show the intersections at b =2,4,6,8 with a maximum error of ~ 0.0013 (at b~2.57) but if we are interested in integrating shapes that are not low order polynomials then we will have some interest in what is happening in other parts of the vector spectrum. The maximum error between the first b=2 node and last b=8 node is dwarfed by the error ~ -0.02026 (at b~34.086) but the largest error by far is ~-0.2844 (at b=0). As explained further down, the reason for the very large error at b=0 is because CGN5 nodes include a centre node.

Assessment Vectors

If we think of CGN's as being power-vector nodes, then we can say that CGN's use the power vector to generate the nodes and the above chart uses the power vector to assess the nodes. Although I am not sure how much of a problem it is that $(Abs(x))^{\wedge b}$ is not a smooth curve for small b-values I will be favouring the use of assessment vectors with the property $V(Abs(x), b)$ is a smooth curve.

It does not matter that the x-nodes and weights were generated from some other unrelated vector; providing the vector is a reasonable vector, then the x-Nodes and weights will make curves that will track the A(b) curve for the unrelated vector. However, using an assessment vector that is different to the Gauss Node vector will mean that we will not know where to expect the intersection points and we will not know how many intersection points to expect.

Instead of the power vector I will sometimes use the Runge vector (with poly-normalised form x^2/(1+b x^2)*(1+b)) as the assessment vector.

Although the Runge vector gets around the power vector problems, it is not always convenient to use. Getting b-values from the area values can be a nuisance. Even going in the easy direction from b-value to the area is not that convenient because we need an expression for those b-values less than zero and another expression for those b-values greater than zero and a series approximation for b-values close to zero.

So, in addition to the Runge vector I also use what I call the IR (integrating rational) vector $(6 b x^2 - 2 (b - 2) x^4)/((b + 1) (b - (b - 2) x^2)^2)$ as the assessment vector. As explained in Appendix A, the IR vector has the same simple relationship between area and shape parameter as the polynomial vector namely that: A=1/(b+1), b=1/A-1 but it generates smooth bell-curves for small b-values. Despite its complex appearance, the IR vector is easy to use.

The Runge Phenomenon

Although all CGN's give a zero error for the y =1 vector, the error curves do not show the y =1 (A=1) vector when the nodes include a centre node. If we think of our vector shape parameter in terms of poly-normalised area A, then immediately adjacent the y =1 (A=1) vector there is the most extreme bell curve $x^2*(1+ 1/\delta)/(1+ 1/\delta*x^2)$ vector with δ ~ zero & A ~1. (Instead of the extreme bell curves we could think in terms of the extreme x^6 power vectors.)

It would be possible to show two values at A=1 on the error curve when our nodes include a centre node, but there is not much value in doing that if we are always going to design our nodes such that the error for the y=1 vector is zero.

These *two adjacent A=1 vectors* may seem to have a similar appearance (except at x=0) but not from the CGN5 perspective because the CGN5 nodes sample the y-value at x=0. When there is a centre node, the b=2 node rather than the b=0 node will be the first curve-fit point relevant to the error curve; hence the huge -0.2844 error at b=0 for the previously discussed CGN5 nodes.

The problem is not so much with the centre node as it is with the polynomial. Although CGN nodes without a centre node (CGN's with an even number of free parameters) do not sample the centre node and therefore the nodes will give the correct A=1 area for the most extreme bell curve, they do not perform that well for moderate bell-curves.

If we think of our vector-shape parameter in terms of A-values instead of the b-values then, although the exclusion of a centre node will mean that the b=0 (A=1) node will be the first curve fit point on the error curve, there will be a large interpolation gap between A=1 (b=0) and A=1/3 (b=2) on the error-curve. In other words, when we adopt A as the vector shape parameter then it will seem that CGN nodes avoid 2/3 of the spectrum.

For appropriately designed non-polynomial vectors (non-CGN nodes) it may be preferable to include a centre node. This is because the A=1 bell-curve is the most extreme vector and therefore it seems reasonable to not insist on zero error for such a vector, whereas we always want a zero error for the y=1 vector (the other A=1 vector) adjacent to the extreme bell-curve vector.

Newton-Cotes Nodes

We can use error curves to analyse and improve the performance of NC nodes and in the process, we will come to understand the performance of NC nodes is about appropriate compromise to achieve a reasonable exponential mapping of the A(b) curve.

NC integration factors are traditionally based on x-nodes spaced at 0,1,2,3,4, etc, but for the purposes of comparison with Gauss-nodes we want to evenly space the x-nodes between 0 and 1. The table below is for 10 RHS x-nodes 0.1, 0.2, 0.3,...0.8, 0.9, 1 plus a centre node weight to be associated with x^0 that can be calculated last. The following weights have been calculated to integrate x^0, x^2, x^4, ...,x^{20}:

x-Node	0.1	0.2	0.3	0.4	0.5
Weight	165.59	-128.15	83.656	-45.418	20.674

x-Node	0.6	0.7	0.8	0.9	1.0
Weight	-7.5421	2.4124	-0.47296	0.22827	0.023651

I will refer to the above NC nodes as NC21 nodes because they have the same number of weights and nodes as the CGN21 nodes. When compared to the CGN21 we can see that the NC21 nodes have the following features:

- **Wildly Fluctuating Weight Values:** We get a weight of about 165.59 for the x=0.1 node and a weight of about -128.15 for the x=0.2 node. The fluctuations between positive and negative weights reduce as we move from 0 to 1, but only the weights associated with x=0.9 and x=1 seem reasonable if we go by CGN standards namely that 'the weights are positive values that add to 1'.

- **Runge Phenomenon:** If we sum all the weights (except the weight for x=0), we get 91.005... Because all the weights including the centre node weight must add to 1 for our nodes to correctly integrate the y=1 line, the centre node weight must be set to -90.005. However, this large weight for the centre node is equal to the error we get when we integrate the most extreme bell-curve vector. So instead of integrating this extreme vector to an area estimate close to 1 the NC nodes will give us a value of 1+ 90.005= 91.005, which means that our NC nodes are shockingly unsuitable for integrating bell-curves. For CGN21 nodes the centre-node weight is 0.1460811336496904 (and half this for the RHS nodes only) which means that the CGN21 nodes will give us an error of 0.073 (below the true value of ~ 1) when integrating extreme bell-curves, which is not great, but it is a huge improvement on the NC nodes.

If we want to improve the performance of these NC21 nodes, then we will need to improve the exponential mapping of the $A(b)=1/(b+1)$ curve. Our 2-parameter vector has the structure $w_k*x_k^{\wedge b}$ where b is the variable and x_k is the predetermined and fixed x-node parameter value and w_k (the weight) is the linear coefficient. Given that we are sticking with the NC x-node spacing our only option for improving the mapping of the $A(b) =1/(b+1)$ curve is by compromising our b-nodes (the points of exact agreement/intersection of our curve fit and the A(b) curve). The following describes a simple procedure that can be implemented with Excel Matrix inversion and the Excel Goal-Seek algorithm to improve the mapping of the A(b) curve:

1. Select two b-values (say $b_1=2$ and $b_2 =10$) and map the 0,1,2,... derivatives at these b-values so that the curve fit will be tangent to the A(b) curve at 2 points. The rows of our matrix will need to express the conditions required to ensure that the derivatives of the exponential curve fit and target curve $A=1/(1+b)$ are the same.

2. Use Excel Goal-Seek to minimise the 'sum of the absolute value of the weights' (SOW) or similar number—the purpose of the SOW number is to allow us to minimise the fluctuation in the weight values and to ensure that they are all positive values. If the sum of the absolute value of the weights is 20 then we might instruct Goal-Seek to reduce the sum of the weights to 10 by varying the b_1 node value. If all goes well, we could repeat the process for some smaller SOW value or we could shift the focus to the b_2 node. Continue the process until the b_1 and b_2 positions have been selected to achieve the lowest SOW value. Each Goal-Seek guess will make Excel perform a matrix inversion and calculate new weights.

3. Identify the x-node with the smallest negative weight and remove this x-node column from the solution. We will also need to pick a condition (row) and remove it so that the number of conditions match the number of linear coefficients (columns).

4. Repeat step 2 then step 3 until all the negative weights are removed.

Using the above process, I eliminated the x=0.2 node then the x=0.5 node and minimised the SOW to get the following NC21.1 weights:

x-Node	0.1	0.2	0.3	0.4	0.5
Weight	0.15824	0	0.19212	0.10464	0

x-Node	0.6	0.7	0.8	0.9	1.0
Weight	0.18565	0.099937	0.048211	0.15019	0.029407

In this solution 0, 1 & 2nd derivative values at b =10.20233 and 0,1,2,3 & 4th derivative values at b=2.69106 were the curve-fit parameters targeted by the matrix linear coefficients (the weights). We will need to add a weight 0.031615 for the x=0 centre node.

This is a fast and easy process that does not require any error measurement. The performance of the conventional NC21 nodes, the NC21.1 nodes and the CGN21 nodes are shown in the chart below. For a complete set of nodes, we need to add the LHS nodes and a centre node with weight of 0.063229.

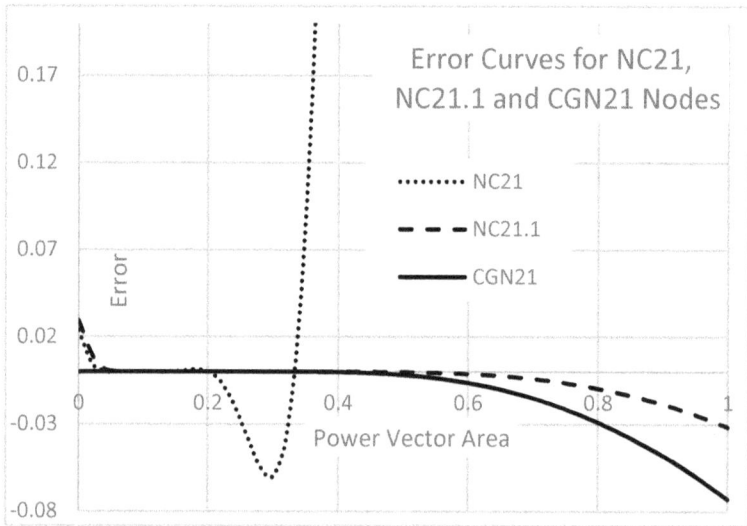

As noted previously, the error measurement (used in the above chart and throughout this document) is equal to 'the estimated-area minus the true-area'. A relative area-error analysis would put more emphasis on errors at the A=0 end of the spectrum.

Noting the scale on the y-axis, in the above chart we can see the Runge phenomenon in the shocking behaviour of the NC21 graph to the right of b=2 (A=0.33...); we do not need a node criterion to tell us that there is a problem with the performance of the conventional NC21 solution. The dip immediately to the left of b=2 on the NC21 error curve is an exaggerated version of the behaviour we see with CGN nodes between b=2 (A=1/3) and b=4 (A=1/5); the dip is not visible on the above CGN error curve because of the scaling of the error axis.

Unlike CGN nodes, the NC21.1 nodes produce a significant error for the zero-area $x^{\wedge\infty}$ vector. This error can be read from the weight (.029) because this is the only weight that has any relevance to the $x^{\wedge\infty}$ vector. CGN nodes do not include an x-node at x=1 therefore they cannot produce an error when integrating the zero-area $x^{\wedge\infty}$ vector. We can fix this error for our NC21.1 nodes by removing the x=1 node and recalculating the weights, however adjacent the A=0 node the CGN's perform much better than the NC21.1 nodes. If we scaled the chart appropriately, we would see that the CGN error adjacent A=0 reaches a maximum value of only about 0.0013 at about b~500 (or A~.002).

NC nodes use the vectors prescribed by the polynomial rule and they use only the weights (linear coefficients) to map the A(b) curve whereas Gauss-nodes use both the exponential vector shape parameter and the weights to map the A(b) curve. Consequently, the Gauss-node exponential provides a better curve fit towards the zero-area U-curves when compared to the modified NC21.1 nodes and a much better fit than conventional NC nodes towards the Runge (A~1) end of the spectrum.

Alternative Vectors

The performance of our nodes will depend on how well our single parameter vectors will map our target curves and how well our two-parameter vectors will map the A(b) curve for the range of b-values appropriate to the target curves.

Although the number of curve fit conditions is important, the performance of the NC21.1 nodes (with only 9 curve-fit conditions) compared to NC21 nodes (with 11) and even CGN21 nodes (with

21) demonstrates the importance of the curve fit conditions. Of course, we know that CGN's are going to control the error to small values in the region of the CGN b-nodes and we know that they are going to integrate low order polynomials exactly, but:

- At what cost? CGN's, particularly CGN's with a centre node, do not map the 1/(b+1) curve well for the Runge end of the spectrum. And although it seems natural to think of the Runge phenomena as being an issue with equally spaced nodes, the NC21.1 error for integrating sharp bell-curves is about half of what we get from CGN21 nodes. This is because the NC21.1 solution compromised the 'exact integration of low order polynomials' in favour of a better overall curve-fit to the A(b) curve.

- And for what benefit? An Excel calculated and graphed error curve using 16-digit CGN21 nodes did not show any intersection nodes on the error curve between about b = 20 and 38. Not that the very small inaccuracies are a concern but that *even if we are only interested in low order polynomials*, then depending on the accuracy of our calculations, the diminishing returns suggests that it may be better to skip a few b-nodes to improve the reach of the nodes and weights rather than follow the polynomial-rule.

We want each weight and each x-node sample point to generate an intersection with the A(b) curve as per the CGN's but after we get to a b-value of say 20 we could opt for an alternative solution and instead of selecting the next node to be 22 we could make it 24. Or rather than thinking in terms of even valued indices we could switch to a power vector and make the next node 23.5. Or, if we want to focus on the Runge end of the spectrum and we are concerned with the fact that low index $(Abs(x))^{\wedge b}$ vectors do not make smooth even curves, we could consider an alternative vector.

An alternative solution might mean that we can escape both the power vector and the polynomial rule. Regardless what $V(x,b)$ vector we use, we will want to find the x-nodes and weights such that for every 'n' number free x-nodes we can satisfy 2n curve-fit conditions for the $A(b)$ curve where $A(b) = \int_0^1 V(x,b)\,dx$.

Finding Gauss-nodes is a problem of finding the two-parameter curve-fits of the form: $w_1*V(x_1,b) + w_2*V(x_2,b) + ... = A(b)$ for $b \sim \{b_1, b_2,... \}$. The weights are always the external linear coefficients in this two-parameter curve-fit. When the single parameter target curve vector changes into the two-parameter vector to map the $A(b)$ curve, the structure of the vector must remain the same—if the x is a base (or index) then it must remain a base (or index) and if b is a base (or index) then it must remain a base (or index). However, when we shift from mapping a $y(x)$ target curve to mapping the $A(b)$ area curve, the 'b' shape parameter becomes a variable, and the x-variable becomes a shape parameter.

This means that when our target curve (single parameter) vector is a polynomial or power vector then we will need the two-parameter $A(b)$ vector to be an exponential. And when our target curve (single parameter) vector is an exponential then we need the two-parameter $A(b)$ vector to be a power vector. But when the vector takes the $f(b*x)$ form (such as $\cos(b*x)$, $\sinh(b*x)$, etc) then apart from some subscripts there will be no change when the roles of the shape parameter and variable are swapped.

Two Kinds of Vectors

The discussion in the following 2 sections includes the development of proof-of-concept alternative Gauss nodes. Because there are so many functions that make suitable vector shapes, the issue of alternative solutions is mostly about finding those 2-parameter vector mappings that allow a mathematical method for finding the linear coefficients and shape parameters. There are two kinds of solutions:

- DE solutions using vectors like cosh(x), cos(x), $e^{\wedge x}$, etc, that use either uniformly spaced data or derivative data at a point, and sometimes include a change of variable substitution.

- Rational partial fraction solutions. Only the Runge rational vector will be included in the alternative rational vector solutions discussed below.

Regardless of whether we are using DE solutions or rational solutions, we will need to find the roots of a polynomial to find the x-nodes. Microsoft Excel should be able to handle the matrix inversions for a small number of nodes and if we put our polynomials in text form, we can drop them into Wolfram Alpha and can quickly get the roots. When the equations are too long for Wolfram Alpha and the calculations require more accuracy than can be provided by Excel then we can use xlPrecision and numerical iteration to get the roots.

5. DE SOLUTIONS

The DE method is a very nice non-iterative mathematical method for finding x-Nodes. And once we have found the x-Nodes we can find the weights (the linear coefficients) by solving linear simultaneous equations.

DE solutions are restricted solutions in the sense that we do not have complete control over which A(b) data goes into the solution. CGN's for example map the b values 0,2,4,6,... We can introduce a new independent variable with a substitution so that the independent variable is not sampled at unit intervals, to get a non-conventional DE solution, but we cannot entirely escape the restrictions. But it may be that a restricted DE solution suits a particular application just as CGN's suit certain applications.

Conventional Difference Equations

The CGN solution is a conventional difference equation solution because the two-parameter CGN vectors are exponential functions. However, cosine and sine functions and hyperbolic cosine and hyperbolic sine functions are also exponential functions.

If we ignore the issue of repeat roots, then conventional difference equations generate vectors of the form $a*x_j^{\wedge b}$ and $x_j^{\wedge b}*(c_1*\cos(a*b) +c_2*\sin(a*b))$ and $(c_1*\cos(a*b) +c_2*\sin(a*b))$. We get the sine and cosine vectors when the roots are complex. And we eliminate the $x_j^{\wedge b}$ term when the base equals 1. (Here the $x_j^{\wedge b}$ is an exponential term not a power term because when we are mapping the A(b) curve, 'b' is the independent variable, A is the dependent variable and 'x' is a parameter.)

CGN DE's produce exponential functions when the roots of the characteristic equations are real. In other situations, the roots of the characteristic equation maybe complex, and the solution will include cosine and sine functions. Depending on the type of linear-coefficient vector we are using to map the target curve, we may want the 2-parameter vector to take the form $x_j{}^{\wedge b}$, $\cosh(b\, x_j)$, $\sinh(b\, x_j)$, $\cos(b\, x_j)$, $\sin(b\, x_j)$, etc. The issue here is not how to produce a curve fit solution to the A(b) curve, but how to produce a curve fit solution that creates a Gauss-node solution. For the curve-fit solution to be a Gauss-node solution, *the 2-parameter vector must be the linear coefficient target vector* with the roles of the shape parameter and independent variable swapped.

If $A(b) = \int_0^1 V(b, x)\, dx$ then we need our DE mapping of A(b) to produce a series of $c_j * V(b, x_j)$ terms. If our single parameter target curve vector is a polynomial, then we will want the 2-parameter solution to be an exponential which means we want the roots of the characteristic equation (CE) to be real. If our single parameter target curve vector is a cosine vector, then we will want the two-parameter vector to also be a cosine vector.

cos(b x) Vector
For cos(b x) as the target curve vector we get: $A(b) = \int_0^1 \cos(b\, x)\, dx = \sin(b)/b$. We need to be able to map the sin(b)/b curve with $w_j * \cos(x_j * b)$ two-parameter vectors. But to get these two-parameter vectors from our DE solution we need to get cosine vectors from the DE solution. This means that we need complex roots with the base of exactly 1 and we need the coefficient of the sine component to be exactly zero.

Because the sin(b)/b function is even about b=0, if we ensure that our DE data is even about b=0 we will eliminate the sine components. The b-values I have used to generate my data are: {-7.5, -6.5, -5.5, -4.5, -3.5, -2.5, -1.5, -.5, +.5, +1.5, +2.5, +3.5, +4.5, +5.5, +6.5, +7.5}. (We can use the symmetry to halve the size of the matrix because exponentials that are even about the y-axis have roots in pairs as: b and 1/b with the CE taking the form $A*b^{\wedge n}$ $+B*b^{\wedge(n-1)} +C*b^{\wedge(n-2)} +...C*b^{\wedge 2} +B*b^{\wedge 1} +A.)$

The CE is: $b^{\wedge}8$ $+1=$ 6.28379526768367*(b^7+b) - 18.6130816700679*(b^6+b^2) +33.7459992527365*(b^5 +b^3) - 40.8373119399985*b^4. This CE has roots: (0.5760866946, ±0.8173885981), (0.708240323, ±0.705971420), (0.873038675, ±0.487650974), (0.98453194, ±0.17520518). The radius generated by these pairs of real and imaginary values is 1 which means there is no exponential term, as required. And the coefficients (weights) associated with the sin(b x) terms are zero as required. The x-Nodes and weights and the error curve for these nodes are given below:

CSN8		CGN8	
x-node	Weight	x-node	Weight
0.176114	0.349113	0.183434642	0.362683783
0.509397	0.310825	0.52553241	0.313706646
0.783794	0.230722	0.796666477	0.222381034
0.956863	0.10934	0.960289856	0.101228536

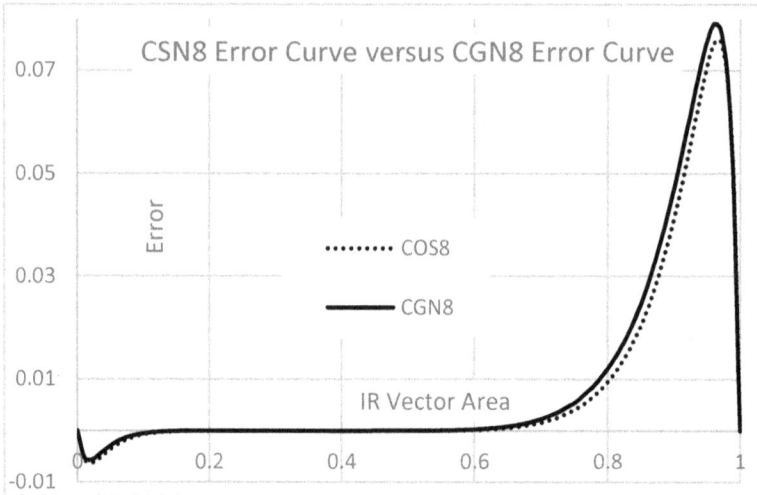

The x-node and weight values are close to the CGN x-node and weight values, and the error curves seem similar.

cosh(b x) Vector

For the cosh(b x) target curve vector we have A(b) = $\int_0^1$ cosh(b x) dx = sinh(b)/b so we need to use w_k*cosh(b*x_k) terms to map the A(b) curve. for b ={-7.5, -6.5, -5.5,...5.5, 6.5, 7.5} we get the DE:

x^8=-1+10.0275010366435*x -41.4355022391717*x^2 +91.7454020940576*x^3 -118.682371798122*x^4 +91.7454020940576*x^5 - 41.4355022391717*x^6 +10.0275010366435*x^7

From this DE we get the following roots and Ln(roots):

Roots	0.38169109	0.44569989	0.58211905	0.82622848
Ln(Roots)	-0.96314367	-0.80810944	-0.54108029	-0.19088393

Roots	1.21031896	1.71786165	2.2436622	2.61991971
Ln(Roots)	0.19088393	0.54108029	0.80810944	0.96314367

CHN8 (cosh(b x) vector)	
x-nodes	weights
0.190883931100000	0.376270576400000
0.541080290800000	0.315331689700000
0.808109442400000	0.214051371000000
0.963143670700000	0.094346362900000

The above x-node and weight values have been rounded to 10 significant figures.

Neither of the above conventional DE solutions using unit b-node spacing seem that interesting as both the cos(b x) nodes and the cosh(b x) nodes generate nodes and error curves that are similar to the CGN nodes and error curves.

Weight-functions

When we are using difference equations, adding a weight-function does not add any complexity to the problem of finding Gauss-nodes; we still need to find the exponential curve fit to an A(b) curve, except that the A(b) area curve will be the integral of 'the vector times the weight-function':

$$A(b) = \int_0^1 V(x,b) \, Wt(x) \, dx$$

Chebyshev Weight-function

Chebyshev nodes of the first kind use the weight-function $Wt(x) = 1/\sqrt{(1-x^2)}$. Such nodes allow us to determine the exact integral of (Low Order Polynomial)/$\sqrt{(1-x^2)}$; increasing the number of nodes allows us to integrate higher order polynomials. According to WolframAlpha:

$A(b) = \int_0^1 x^b/\sqrt{(1 - x^2)}\, dx = sqrt(\pi)\ \Gamma((1+b)/2)/(2\ \Gamma(1+b/2))$, where Γ designates the Gamma function.

When we use difference equations to map the above area curve with an exponential curve-fit to b =2,4,6,8 then we get:

0.628318530717956*0.951056516295154^b
+0.628318530717961*0.587785252292474^b

We need to add a centre node so that the nodes integrate $x^0/\sqrt{(1 - x^2)}$ to give a total of 5 b-nodes The nice thing about the linear coefficients (the weights) is that they are all the same (equal to pi/5 in this example). And the nice thing about the x-nodes is that they can be generated by the simple formula: x_k =cos((2 k -1)/(2 n)*pi) for k =1,2,3,..n.

If we ignore the above solution apart from the x-nodes we can generate the Chebyshev polynomial of the first kind: 16*(x^2-0.951056516295154^2) *(x^2- 0.587785252292474^2)*x that passes though the above x-nodes and the centre node, we will get a 'wave-curve' with 5 roots, and 6 extremes (2 edge values, 2 troughs and 2 peaks). Multiplying it by 16 normalises it so that the extremes will be equal to ±1:

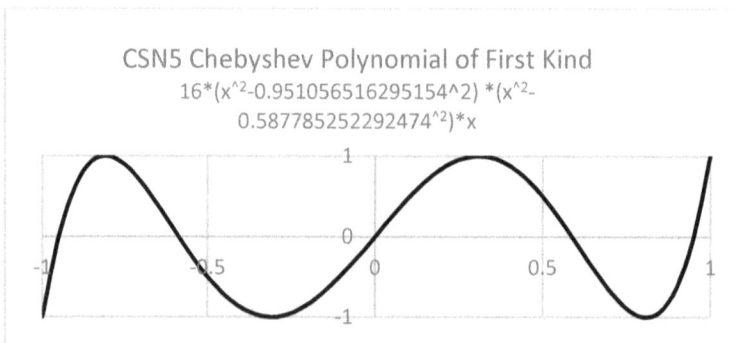

CSN5 Chebyshev Polynomial of First Kind
16*(x^2-0.951056516295154^2) *(x^2-
0.587785252292474^2)*x

The wave-curve for equally spaced x-nodes would show extremes near the edges larger than the extremes near the centre, just as when we map the Runge curve with equally spaced x-nodes, the absolute value of the in-between errors will be larger at the edges. And like Runge's phenomenon this behaviour amplifies as we increase the number of equally spaced x-nodes. See Wikipedia: Chebyshev Nodes.

We can easily get weights for the above nodes the same way we get weights for Newton-Cotes nodes. And although these nodes will not be Gauss nodes, according to the Wikipedia: Clenshaw Curtis Quadrature article, Clenshaw Curtis integration, has 'fast-converging accuracy comparable to Gaussian quadrature rules'. So, for the sake of a little insight into what we want from our error curves and a Gauss node criterion, it may help to compare the Chebyshev nodes with CGN nodes.

Using the rule $x_k = \cos((2k-1)/(2n)*pi)$ with $n=16$ we can derive the following CSN16 nodes and weights:

CSN16 x-Nodes	Weights
0.995184726672197	0.0168027552298682
0.956940335732209	0.0583364640711181
0.881921264348355	0.0917183132017243
0.773010453362737	0.125129618162390
0.634393284163645	0.151392461635083
0.471396736825998	0.173419411772713
0.290284677254462	0.187749728114440
0.0980171403295608	0.195451247812585

The error chart below compares the CSN16 nodes with the CGN16 nodes using the IR assessment vector.

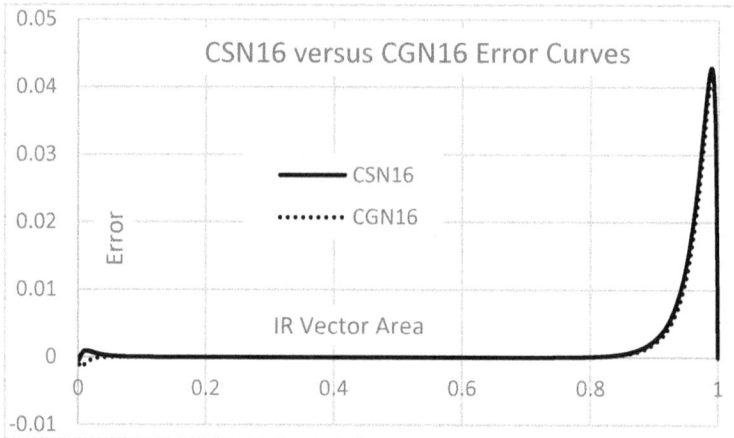

The overall performance of both sets of nodes seem quite similar with both suffering from a significant Runge-curve deficiency. As shown on the chart below, the CSN16 nodes generate an unexpected node near b= 271.7682 for the IR assessment vector (& near b= 390.0015 for a power assessment vector).

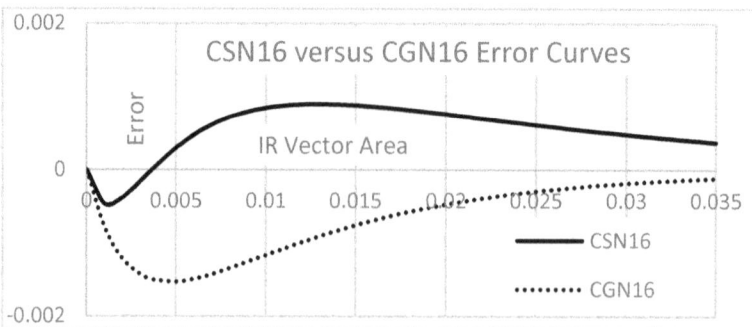

Even if the Chebyshev nodes always give an extra node in the small area part of the error curve the Wikipedia reported performance of the CSN nodes seems surprising. When measured with the IR assessment vector the CSN16 performance is better than the CGN16 performance only for IR vectors with A-values from 0.001 to 0.013, 0.695 to 0.760 and 0.996 to 0.999 and except very close to the additional node at A~.003666, the ratio of the (CGN error)/(CSN error) is always smaller than 10. By contrast there is a substantial spectrum of vectors for which the CGN performance (as measured by the (CSN error)/(CGN error) using the IR assessment vector) is more than 100 times better than the CSN performance.

Apparently then the high-performance regions of CGN nodes are not always important. So why not try and improve the bell-curve end of the spectrum where both CSN and CGN nodes fail badly.

Substitutions

We can also implement a change of independent variable to adapt the DE method to non-DE vectors without the need for any new theory. We can use conventional exponential difference equations to find the solution in terms of some new independent variable B which is defined by the substitution $B = f(b)$. We need $B = f(b)$ to be such that when we substitute $f(b)$ into the 2-parameter exponential mapping $a_1 * x_1^{\wedge B} + a_2 * x_2^{\wedge B} + a_3 * x_3^{\wedge B} \ldots$ our two-parameter vector takes on the required form. This means we determine the substitution $B = f(b)$ first and select our data input according to the $b = f^{-1}(B)$.

By ensuring that the independent variable B is sampled at one unit spacing we can use conventional difference equation methodology to get an exponential in the B parameter and then substitute $B = f(b)$ to get the parameter we want. Even the CGN DE solution required a substitution because the independent variable skips the odd polynomials.

$b^{\wedge x}$ Vector

With the $b_j^{\wedge x}$ exponential vector as the target curve vector then the two-parameter vector $w_j*b^{\wedge x_j}$ will be a power curve vector. Therefore, we need a change of independent variable if we want to use a DE vector and conventional DE methodology to map: $A(b) = \int_0^1 b^{\wedge x}\, dx = (b - 1)/Ln(b)$ with two-parameter power vectors.

If we suppose we have found a conventional exponential solution $\sum c_j*x_j^{\wedge B} \sim A(b)$, where $B=f(b)$, then we first want to find the back-substitution $B=f(b)$ such that the exponential terms convert into a power series terms. And if we put $B =Log_k(b)$ then we get $x_j^{\wedge B} = x_j^{\wedge Log_k(b)}$. But: $x_j^{\wedge Log_k(b)} = b^{\wedge Log_k(x_j)}$ which gives us the power vector that we need.

To make the back substitution $B = Log_k(b)$ valid we need to sample b using $b= k^{\wedge B}$ with B sampled at one unit spacing. For k values less than about 4.17835 I got complex roots, so I chose k=5. The following power series maps $(b-1)/ln(b)$ at b ~ 1, 5, 25, 125, 625, 3125, 15625, 78125:

$0.105311986561844*b^{\wedge 0.0391596586155792}$ $+0.185074613091047*b^{\wedge 0.19983382752827}$
$+0.193520644643058*b^{\wedge 0.381831177045866}$ $+0.190163819675945*b^{\wedge 0.58579113682368}$
$+0.125902278421108*b^{\wedge 0.73750503137545}$ $+0.111870063223023*b^{\wedge 0.859905849580715}$
$+0.0600405668049007*b^{\wedge 0.945119096841013}$ $+0.0281160168612189*b^{\wedge 0.98881786121762}$

As always, the weights are the linear coefficients, but in this case the x-nodes are the indices. The x- node and weight values are:

Exponential b^x vector (EXN16)	
x-nodes	weights
3.91596586155790E-02	1.05311986561844E-01
1.99833827528270E-01	1.85074613091047E-01
3.81831177045866E-01	1.93520644643058E-01
5.85791136823680E-01	1.90163819675945E-01
7.37505031375450E-01	1.25902278421108E-01
8.59905849580715E-01	1.11870063223023E-01
9.45119096841013E-01	6.00405668049010E-02
9.88817861217620E-01	2.81160168612190E-02

The charts below compare the integration error of the CGN16 nodes versus the EXN16 using the IR vector as the assessment vector:

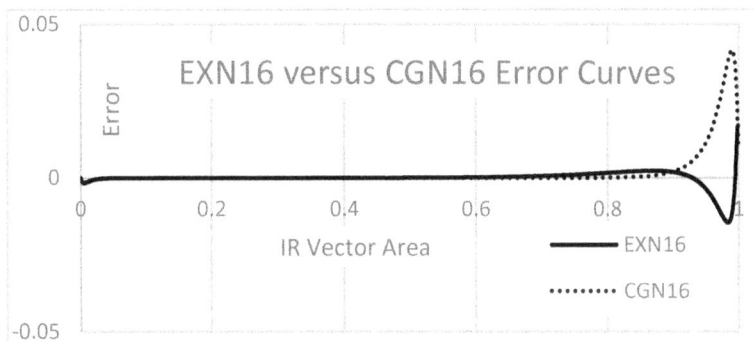

For small area U-curve vectors the EXN16 and the CGN16 seem to have a similar performance. For large area vectors (sharp bell curves) the EXN16 nodes perform better than the CGN16 nodes.

The EXN16 nodes have a very small negative error at A=1, but that can only be due to calculation error as the b=1 value was included in the b-nodes. This error is not visible on the above chart but if we sum the weights, we get a value of about 0.999999989. We could add a centre node to fix the error, or we could fix it with xlPrecision. Although the EXP16 nodes have a focus on the Runge end of the chart with one Gauss node at 0.9945 and another at 0.9273, it is interesting to note that their performance falls short of the CGN nodes for moderate Runge vectors.

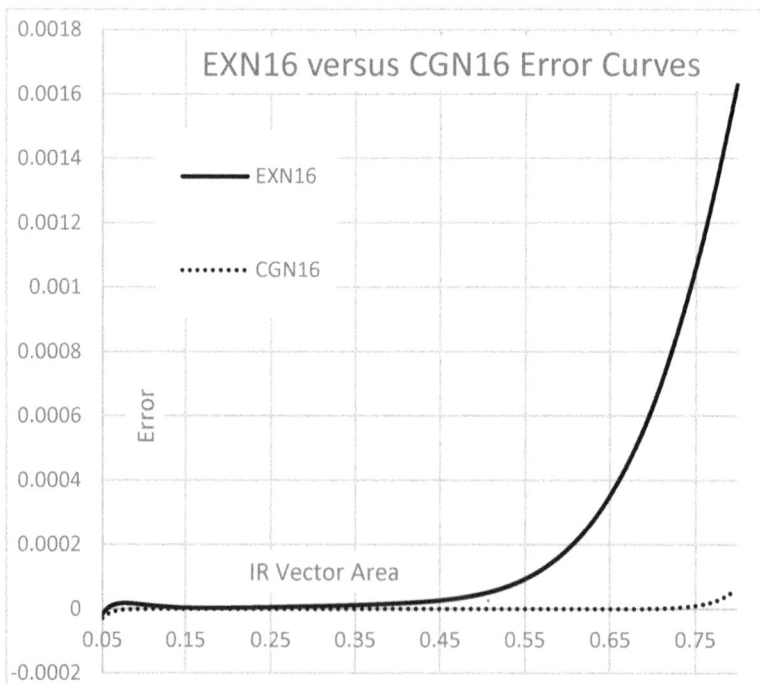

EXN16 versus CGN16 Error Curves

Tangent Curve-Fits

Instead of intersecting the A(b) curve at pre-selected b-nodes, we can create tangent curve fits that map the derivatives of the A(b) curve at a pre-selected point on the A(b) curve.

The Newton Cotes NC21.1 example from above manipulated the vector linear coefficients to create tangents to the A(b) curve, but for valid Gauss nodes we need to manipulate both the external linear-coefficient and the vector-shape-parameter to construct the curve-fit and the number of curve-fit conditions satisfied must not be less than the number of free parameters. This means we need to covert a Taylor series expansion (TEX) into a differential equation.

Suppose that we have found the TEX to some A(b) function at some point. We then convert the TEX series into derivatives up to order 'n' and list them in a excel column and then find the difference equation coefficients that correspond to the column of values. But because we want to use an exponential to map derivative values, instead of working with $a*b^{\wedge x}$ vectors it will be more convenient to work with $a*e^{\wedge(b\,x)}$ vectors.

In the case of mapping the derivatives at a single point we are saying we are interested in behaviour of the Gauss node curve-fit near some b-value and maybe not so interested in the curve-fit remote from the b-value. This could be the case if we are interested in dividing up our curve into small segments and zooming into each of the segments by stretching them out to x=-1 to x=1 and applying our Gauss nodes to the stretched-out segment. If a curve has the TEX: $a +b*x +c*x^{\wedge2} +d*x^{\wedge3}+...$ and we then stretch out the portion of the curve from x= -1/10 to +1/10 to x= -1 to x= +1 we will get the TEX: $a+ 1/10*b*x +1/100*c*x^{\wedge2} +1/1000*d*x^{\wedge3}+...$ Because the odd terms will be handled by the node symmetry, after the y=1 vector the next most important vector will be $y=x^{\wedge 2}$. The three sets of Gauss nodes developed below are tangent curve-fit solutions for:

1. The power vector area curve $A(b) = \int x^{\wedge b}\,dx$ curve at $b=2$.

2. The hyperbolic cosine area curve $A(b) = \int \cosh(b*x)\,dx$ at $b=0$.

3. The exponential area curve $A(b) = \int b^{\wedge x}\,dx$ at $b=1$.

A single error chart for these 3 sets of nodes is given at the end of this sub-section.

$x^{\wedge b}$ Vector (@ b=2)

With the target curve vector $x^{\wedge b}$ and with $A(b) = \int_0^1 x^{\wedge b}\,dx = 1/(b+1)$, we want the area curve mapping to be in $w_k * x_k^{\wedge b}$ terms. Instead of mapping the area curve at $b=0,2,4,...$ (as per CGN's)—here we want the exponential mapping to generate the correct lowest order derivatives at $b=2$.

TEX$_{@b=2}$: $1/3^{\wedge 1} - 1/3^{\wedge 2}*(b - 2) + 1/3^{\wedge 3}*(b - 2)^{\wedge 2} - 1/3^{\wedge 4}*(b - 2)^{\wedge 3} + 1/3^{\wedge 5}*(b - 2)^{\wedge 4}$

The kth derivative value follows the rule: $1/3^{\wedge (k+1)}*(-1)^{\wedge k}*k!$. If we list these derivative values in a column of 16 values then we will get the DE: $x^{\wedge 8}=-64/3*x^{\wedge 7} -1568/9*x^{\wedge 6} -18816/27*x^{\wedge 5} -117600/81*x^{\wedge 4} - 376320/243*x^{\wedge 3} -564480/729*x^{\wedge 2} -322560/2187*x -40320/6561$, with roots: $\{r_1{\sim}{-}7.6210439122964213686,\ r_2{\sim}{-}5.2468928804260015260,\ r_3{\sim}{-}3.5861720033936650747,\ etc\}$

The weights will be the linear coefficients that make the curve-fit exponential generate the correct derivative values at $b=2$. The derivative value for the kth order derivative of each of the $e^{\wedge (rj*b)}$ vectors is given by $e^{\wedge (2*r)}*r^{\wedge k}$. So, if we generate a matrix of values with one column for each root value and one row for each derivative, we can solve the simultaneous equations to find the weights required to generate the required derivative values. We will get the x-node values as $x_j = e^{\wedge r_j}$ when we put the $e^{\wedge(r_j*b)}$ vector into $x_j^{\wedge b}$ target vector form. We need to include a centre node because we want to integrate y= 1 vector exactly, but we can tell from the small value of the centre node weight that these nodes will be function well for extreme bell-curves.

PTN17	
x-nodes	Weights
0.000000000000000000	0.000040103147498360
0.000490030015655246	0.001454772723902930
0.005263848422844710	0.010208502988054200
0.027704179203580800	0.039419126965305400
0.095499417200068600	0.102137929095767000
0.241174777595664000	0.191084762102757000
0.472195487855597000	0.262809805790833000
0.739904669857212000	0.254988252786668000
0.944820914951593000	0.137856744399214000

Cosh(b x) Vector (@ b =0)

$\int_0^1 \cosh(b\,x)\,dx = \sinh(b)/b$ and the even derivative values at b=0 for the kth order derivative of $\sinh(b)/b$ are given by the formula $1/(k+1)$. We will get the same values from the $1/(n+1)$ formula for the area under the $x^{\wedge n}$ polynomial vector from x=0 to 1 when n =k. Therefore, CGN nodes are the solution to this problem because CGN nodes map the $1/(n+1)$ curve at even values of n. This is to say that, using the CGN8 node values:

- We can write a curve-fit that intersects the $1/(b+1)$ curve at b =0,2,4,...12,14:

$$0.362683783378362*0.183434642495649\text{^}b$$
$$+0.313706645877887*0.525532409916329\text{^}b$$
$$+0.222381034453374*0.796666477413626\text{^}b$$
$$+0.101228536290376*0.960289856497536\text{^}b$$

- and using the same coefficients we can write a curve-fit that creates a tangent to sinh(b)/b at b=0:

$$0.362683783378362*\cosh(0.183434642495649*b)$$
$$+0.313706645877887*\cosh(0.525532409916329*b)$$
$$+0.222381034453374*\cosh(0.796666477413626*b)$$
$$+0.101228536290376*\cosh(0.960289856497536*b)$$

$b^{\wedge x}$ *Vector (@ b=1)*

When b=1, $b^{\wedge x}$ creates the y=1 vector and Gauss nodes that create a tangent to $1^{\wedge x}$. $\int_0^1 b^{\wedge x}\,dx$ = (b - 1)/log(b). However, we need a power expression not an exponential expression, so we put: b $=e^{\wedge B}$ in (b - 1)/log(b) to get: $(e^{\wedge B}$ - 1)/log($e^{\wedge B}$) = $(e^{\wedge B}$ - 1)/B which has TEX:

$$1 + B/2 + B^{\wedge}2/6 + B^{\wedge}3/24 + B^{\wedge}4/120 + B^{\wedge}5/720 + ...B^{\wedge k}/(k+1)!...$$

We now need to:

1. find the exponential series in index variable B that maps the derivatives of this TEX and then

2. substitute Ln(b) for B to convert the exponential terms into constant*$e^{\wedge(constant*ln(b))}$ and then

3. use the fact that $e^{\wedge(c*ln(b))}$ = $(e^{\wedge ln(b)})^{\wedge c}$ = $b^{\wedge c}$ to get the power term expression.

For example, if we map the first 8 derivative values with 4 exponential terms we get:

1. $0.17392742256924*e^{\wedge}(0.06943184420294*B)$
   $+0.326072577430336*e^{\wedge}(0.3300094782081*B)$
   $+0.326072577432494*e^{\wedge}(0.669990521792*B)$
   $+0.17392742256793*e^{\wedge}(0.930568155798*B)$

2. $0.17392742256924*e^{\wedge}(0.06943184420294*Ln(b))$
   $+0.326072577430336*e^{\wedge}(0.3300094782081*Ln(b))$

$$+0.326072577432494*e\wedge(0.669990521792*Ln(b))$$
$$+0.17392742256793*e\wedge(0.930568155798*Ln(b))$$

3. $0.17392742256924*b\wedge(0.06943184420294)$
$$+0.326072577430336*b\wedge(0.3300094782081)$$
$$+0.326072577432494*b\wedge(0.669990521792)$$
$$+0.17392742256793*b\wedge(0.930568155798)$$

We now have a power series that maps the derivatives of (b - 1)/log(b) @ b=1. The x-nodes and weights are:

x-Nodes	Weights
0.06943184420294	0.17392742256924
0.3300094782081	0.326072577430336
0.669990521792	0.326072577432494
0.930568155798	0.17392742256793

There are some interesting patterns in the above values, namely: $w_1\sim w_4$, $w_2\sim w_3$, $x_1+x_4\sim 1$, $x_2+x_3\sim 1$. Furthermore, the error curve for these nodes and weights has zeros at b=0,1,2,3,4,5,6,7 when using the power vector as the assessment vector. This means the combined LHS and RHS nodes and weights will integrate $(Abs(x))\wedge b$ exactly for integer values of b from 0 to 7. And this also means we can generate these nodes the same way we generate CGN nodes. The XTN16 nodes are given below.

XTN16	
x-Nodes	Weights
0.019855072	0.050614268
0.101666761	0.111190517
0.237233795	0.156853323
0.408282679	0.181341892
0.591717321	0.181341892
0.762766205	0.156853323
0.898333239	0.111190517
0.980144928	0.050614268

Error Charts for the Tangent Nodes

The chart below compares the above tangent solutions over the entire spectrum:

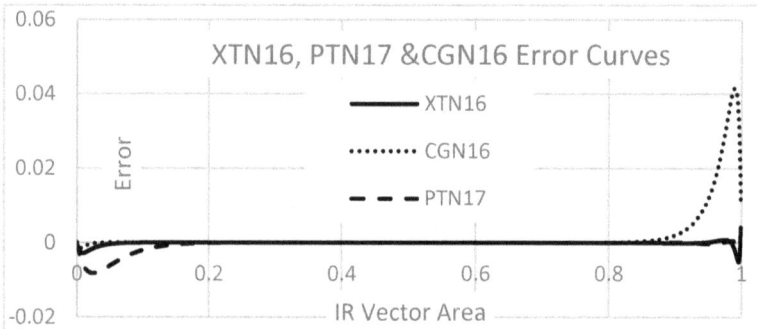

We know from the small value of the PTN17 centre-node weight that the PTN17 nodes are going to be good at integrating extreme bell-curves, and this is confirmed by the above chart, but they are not so good at integrating U-curve vectors. The above chart shows us that XTN nodes will perform much better than the CGN17 nodes for bell-curve vectors and better than the PTN17 nodes for U-curve vectors. Because the above chart covers the entire spectrum it conceals the large differences in performance across the spectrum of small errors—for areas below about 0.5 CGN's perform significantly better than the XTN nodes.

6. RATIONAL SOLUTIONS

Runge Vectors

Instead of the $1/(1+b\,x)$ rational I have adopted the inverted bell-curve: $x^2/(1+b\,x^2)$. I call this vector the Runge vector but my version has been inverted to make it look more like a polynomial vector: $x^2/(1+b\,x^2) = 1/b*(1 - 1/(1+b\,x^2))$. Multiplying it by $(1+b)$ to get $(1+b)\,x^2/(1+b\,x^2)$ poly-normalises the vector.

The chart below shows a range of poly-normalised $(1+b)*x^2/(1+b\,x^2)$ Runge vectors with areas 0.05, 0.2, 0.4, 0,6, 0.8, 0.95.

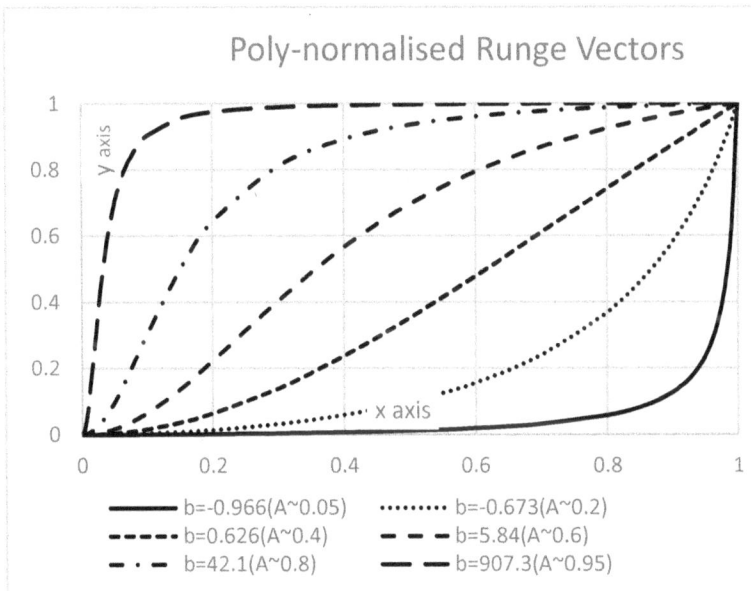

Poly-normalised Runge Vectors

b=-0.966(A~0.05)   b=-0.673(A~0.2)
b=0.626(A~0.4)   b=5.84(A~0.6)
b=42.1(A~0.8)   b=907.3(A~0.95)

Although all the vectors are smooth curves because these shapes are even about the origin, the A~0.95 vector, has a relatively sharp curvature at the origin (y=0, A=0) which is not clearly visible on the above chart.

If we want to create two Gauss x-nodes and two corresponding weights with the Runge vector, then we want:

$w_1 x_1^{\wedge 2}/(1+ b\ x_1^{\wedge 2}) + w_2 x_2^{\wedge 2}/(1+ b\ x_2^{\wedge 2}) \sim A(b)/(1+b)$ (exact @ b_1, b_2, b_3 and b_4)

We are looking to find the values of the coefficients x_1, x_2, w_1, w_2 to achieve a curve-fit at the b-nodes b_1, b_2, b_3, b_4. Using N_0 and N_1 for the unknown numerator coefficients and D_1 and D_2 for the unknown denominator coefficients, we can write out the combined rational as: $(N_0 + N_1*b)/(1+ D_1*b + D_2*b^{\wedge 2})$ and find the coefficients to track $A(b)/(1+b)$. We want this expression to be exact for b $=b_1$, b_2, b_3, b_4. We can linearise this expression by multiplying both sides by the LHS denominator and then taking all the unknown-coefficient terms back to the LHS by subtraction to get:

$N_0 + N_1*b - D_1*b*A(b)/(1+b) - D_2*b^{\wedge 2}*A(b)/(1+b) \sim A(b)/(1+b)$

There are 4 expressions which we want to be exact for the four b-node values and so we use matrix inversion to solve the simultaneous equations to find the values of the coefficients N_0, N_1, D_1 and D_2.

And after we have found the combined rational coefficients, we then need to find the roots of the denominator to get the x-nodes. For the 4 b-nodes we will have two vectors, two x-values x_1 and x_2 and two weights w_1 and w_2 which will give us a rational of the form: Numerator/$((1+ b\ x_1^{\wedge 2})\ (1+ b\ x_2^{\wedge 2})) \sim A(b)/(1+b)$. If we factor $x_1^{\wedge 2}$ *$x_2^{\wedge 2}$ out of the above denominator we can get the b-roots r_1 and r_2:

$((b\ x_1^{\wedge 2} +1)\ (b\ x_2^{\wedge 2} +1)) = x_1^{\wedge 2}\ x_2^{\wedge 2}\ ((b +1/x_1^{\wedge 2})\ (b +1/x_2^{\wedge 2}))$ $=$constant*$(b -r_1)(b -r_2)$ which means that $r_1 = -1/x_1^{\wedge 2}$, $r_2 = -1/x_2^{\wedge 2}$, or that:

- $x_1 = $ sqrt$(- 1/r_1)$, $x_2 = $ sqrt$(- 1/r_2)$, etc

These roots are the denominator roots, not the b-nodes where the rational intersects the $A(b)/(1+b)$ curve. This means that after we have found the roots of the denominator, we can then immediately get the x-nodes for each vector.

Like the DE solution once we have found the vector shape parameters (the x-nodes) we can put the vector shape parameter values back into the vectors and determine the linear coefficients (the weights) required to fit the b-variable vectors to the curve-fit points by solving some linear simultaneous equations.

The above procedure is the conventional procedure used for fitting a rational to data. The only issue we need to keep in mind for Runge Gauss nodes is that we need to ensure that the order of the numerator is one less than the order of the denominator, so that the $w_k x_k^{\wedge 2}/(1+ b x_k^{\wedge 2})$ partial fraction vectors have order 0/1 (in variable b).

Gauss nodes are about using a two-parameter vector $V(x,b)$ to map the vector $A(b)$ curve where $A(b)$ is the integral of the $V(x, b)$ dx from x=0 to 1. The rational solution and the DE solution provide different methods for generating curve fits to the corresponding $A(b)$ curves.

Newton-Cotes nodes restrict the x-nodes, and they also restrict the vector shape parameters to follow the polynomial rule. Traditional Gauss-node vector shape parameters (including CGN's) are restricted by the polynomial rule and other non-traditional Gauss-nodes (discussed above) that we find using DE's, also have the vector shape parameters restricted by the DE method. However, the rational solution allows us to choose the vector shape parameter without restriction so that we can choose any points of intersection on the A(b) curve. Every choice in the range of b=-1 to b = infinity will generate valid Gauss nodes.

For any selection of b-values we can easily find a rational to map the A(b) curve or other target data and then find the roots of the denominator so that we can split the rational into its partial fractions (which become the vectors) and we can do that without any restrictions on the spacing of the b-nodes.

Of course, we can choose any shape parameters we like for the DE vectors too, but we will then have to use multi-variable iteration to find the x-nodes and weights. Furthermore, unlike the DE solution the rational solution only requires basic maths. However, the Runge area formula is not as convenient as many of those area formulas for the DE solutions.

Runge Vector Area

Using A(b) to represent the poly-normalised area from x=0 to x=1 of the Runge vector then: $\int_0^1 x^2/(1+ b x^2) = A(b)/(1+b) =$

- $+1/b -atan(\sqrt{b})/ b^{(3/2)}$,   or

- $-1* (1/(-b) -atanh(\sqrt{(-b)})/(-b)^{(3/2)})$

When b > 0 we can use the first expression and when b < 0 we can use the second expression.

And when A(b) is close to zero we can use a Taylor expansion.

We may also need to be able to go the other way and get the Runge b-values given the poly-normalised area. For a first guess I have used:

$b \sim b_g = (3*A-1)/(1-A)^2*(1 +6*A +2.797*A^2)/(1 +5.433*A +1.506*A^2)$

For accurate b-values, we can start with the above first guess and then use the Newton method to iterate our way towards an accurate value.

Suppose we are using multi-variable iteration to minimise some error; that means we want to increment the b-values (shape parameters) to minimise the error, but we do not really care about the b-values. Instead of incrementing the b-values with the multi-variable iteration process we can increment the B-value defined by the equations:

$$b = (3*B-1)/(1-B)^{\wedge 2}$$

$$B = (2*b - SQRT(8*b + 9) + 3)/(2*b)$$

The B-value is a convenient Runge vector shape parameter. The advantage of the B-value is that like the A-value it is scaled from 0 to 1 rather than from -1 to infinity, and when switching between the B-value and b-value, iteration and approximation is not required. The B-value corresponds exactly with the A-value at b = {-1,0,∞}.

Gauss-node Criteria

The fact that the Runge vector allows us to select any b-nodes we want and then find the solution without iteration, means that for a fixed number of b-nodes (say 4) there is an unlimited number of Runge Gauss node solutions. Instead of CGN4 we will have RUN4.critera 1, RUN4.criteria 2, ... A criterion could be a list of all the b-nodes, or it could be some error criteria. The polynomial rule is the hidden criteria for the CGN nodes.

Predetermined b-nodes

The Runge vector makes it possible to choose the b-values where our two-parameter curve-fit intersects the A(b) curve (the b-nodes). If we are of the opinion that poly-normalised area is a better shape parameter than the Runge b-value then we could space the b-nodes at equal area increments and then find the Gauss-nodes to match.

The Chart of 'Poly-normalised Runge Vectors' given above is almost a chart of equal area increment vectors with an area increment of 0.2, except that the first and the last vectors have areas of 0.05 and .95 (instead of 0 and 1) with increments of 0.05 to the adjacent boundary vectors. We want to see what we get when we turn these b-values into Gauss-nodes.

There are 6 vectors and 6 b-node targets which means we are looking to find 6 curve-fit coefficients (3 x-nodes and 3 weights) to construct the curve-fit. To fully specify the nodes, we can either use b-nodes= {-0.966, -0.673, 0.626, 5.84, 42.1, 907.3} or A-nodes = {0.95, 0.8, 0.6, 0.4, 0.2, 0.05}. But because we do not have a A=1 node we will need a centre node and weight to ensure we can integrate y= constant.

Runge7, A={0.95, 0.8, 0.6, 0.4, 0.2, 0.05 }	
x-Nodes	Weights
0.0000000000	0.0366490769
0.1586915595	0.3021018858
0.6012818358	0.5089138224
0.9554027965	0.1523352149

The charts below compare these nodes with CGN nodes for a Runge, IR and Polynomial assessment vectors. (In theory we should be comparing these Runge nodes with CGN7 nodes except that the CGN7 nodes have a centre node with a weight of 0.208979592*2 producing a very large error near A=1. The Runge7 nodes avoid some of the centre node problem, because the largest Runge node has A ~ 0.95 which is reasonably close to A=1 whereas the largest CGN7 node is only A~0.33333)

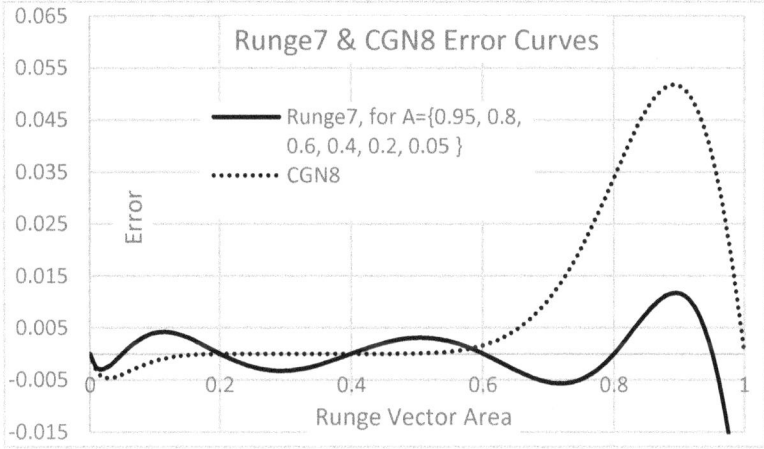

Runge7 & CGN8 Error Curves

Runge7, for A={0.95, 0.8, 0.6, 0.4, 0.2, 0.05 }

CGN8

Error

Runge Vector Area

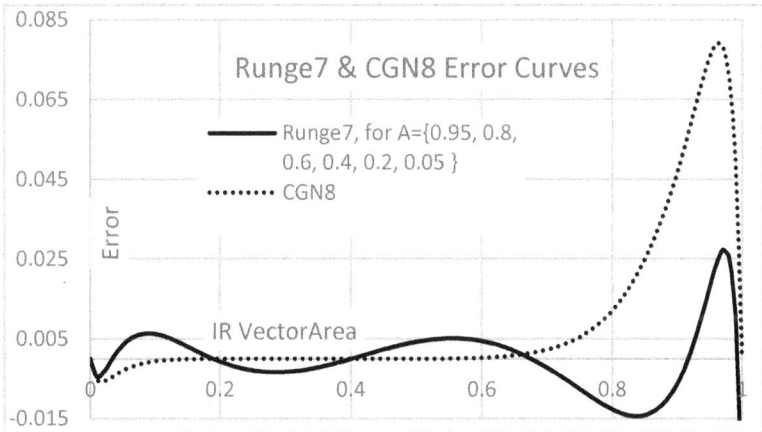

Runge7 & CGN8 Error Curves

Runge7, for A={0.95, 0.8, 0.6, 0.4, 0.2, 0.05 }

CGN8

Error

IR VectorArea

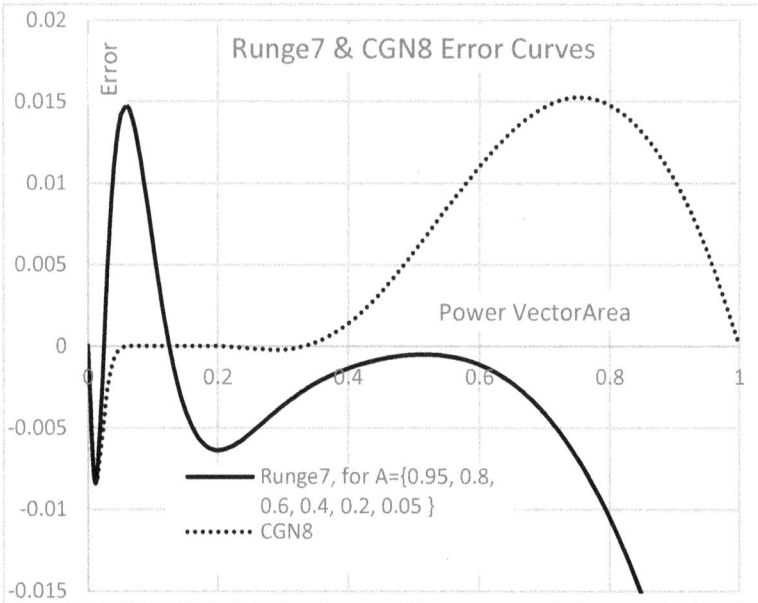

As we can see in charts above, even CGN's without a centre node appear to show a somewhat deficient performance towards the Runge end of the spectrum despite the perfect performance for the most extreme bell curve with A=1. When assessed with the power vector, the Runge nodes with a centre node do not perform so well for the most extreme bell curve despite the A=0.95 node being so close to the A=1 node.

There is nothing special about the selection of Runge vectors (A-node shape parameters) used to create the above charts—any selection of distinct A-nodes will generate Gauss nodes. And given the arbitrary nature of our node selections and the fact that we do not know what we want, for the sake of convenience it makes more sense to use B-nodes instead of A-nodes.

The importance of the Runge vector is that not that any selection of vector shape parameters will give us Gauss nodes because that is the case for all vectors. The importance of the Runge vector is that we can choose any selection of vector shape parameters we like (such as equal area increment vectors) and find the corresponding x-nodes and weights without the need for multi-variable iteration. And the Runge vector solution is a purely algebraic solution.

If we were looking to develop the above Runge7 nodes, the next step in this process would be to make some adjustments to the A-nodes (or B-nodes) based on our judgements about the error curves. We could manually shift our shape parameters around till we get the kind of Gauss node error curve that we want, without the need for multi-variable iteration, except that we might not know what we want. So instead, we can attempt to develop nodes based on some error criteria which means we will need to use multi-variable iteration.

7. ITERATIVE SOLUTIONS

Suppose we want to find the nodes that achieve some error condition measured across the entire spectrum. Because we do not know how the error values relate to the parameter values (the independent iteration variables) we are manipulating, at each step we will need to measure the error values and attempt to determine the direction for the next increment. In this situation we will need to use a multi-variable iteration process and attempt slowly step our way towards the minimum error condition.

We can use the x-nodes and weights as the independent iteration variables. As explained in Appendix D, when we are iterating x-nodes and weights then we will first need to find an initial set of x-nodes and weights that generate the required number of intersections with the A(b) curve and limit the size of our incremental changes in the x-node values and weight values as required to ensure we do not to lose any intersections.

Even though we can use any vector when iterating the x-nodes and weights, I will be sticking with the Runge vector. And when using vector shape parameters as the fundamental independent variables then I will typically be using the Runge vector B-nodes as the fundamental independent iterative variable.

Full Spectrum FEN Nodes

Flat-error nodes (FEN's) are x-nodes and weights that ensure the maximum absolute value of the errors (the in-between errors at the peaks and troughs on the error curve) is constant across some spectrum of vectors. Full Spectrum FEN nodes are nodes that span the entire spectrum of vectors from A= 0 to 1. The error curve for these nodes looks a little like the Chebyshev polynomial of the first kind.

Full spectrum FEN nodes will maintain a moderate performance across the entire range of vector shapes. Each set of FEN nodes comes with a maximum error value that limits how badly we can expect the FEN nodes to fail for the most difficult of Runge vectors. Full spectrum FEN nodes are extreme nodes that are typically not going to be suitable for zooming into difficult parts of a target curve.

Finding FEN's requires that we find the positions of the extreme errors between the nodes, measure the value of the error, and then determine the rate of variation of such errors with respect to the independent parameters. We need the rate of variation of the error-variable (the dependent variable) to be small enough so that it seems approximately linear with respect to the small changes in the independent variables so we can use linear methods to generate the next-guess.

We can use the B-value defined in the above discussion of 'Runge Vector Area' and the position of the maximum error can be found using a quadratic. Or we can iterate the x-nodes and weights as the independent variables as explained in Appendix D.

One of the interesting aspects of these full spectrum FEN nodes is that they escape the need for nomination of any arbitrary criterion other than the choice of the error definition and the choice of the vector.

When the FEN nodes include a centre x-node then the weight value associated with the centre node will equal the maximum error value for the nodes. This is the error we will get when we integrate a Runge vector that has a b-value that aligns with a peak or a trough on the error curve. The following x-nodes and weights and error curves are the full-spectrum Runge FEN9 x-nodes and weights.

FEN9	
x-Nodes	Weights
0.0000000000000000	0.0031518074279546
0.0260546782407331	0.0652035860072439
0.1850187750545720	0.2891675660339270
0.6134988623765920	0.4949154596376980
0.9557080404232400	0.1475615808931750

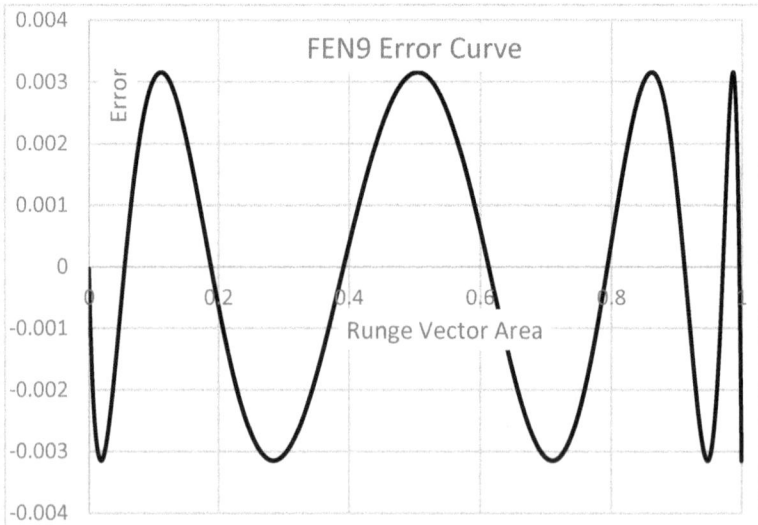

FEN9 Error Curve

The chart below continues the practice of comparing alternative nodes with CGN's with an even number of free parameters (10 in this case). When our nodes exclude a centre node then they will always include an A=1 node (just as CGN's with an even count of free parameters always include and A=1 node) because that is the only way we can make them integrate y=1 exactly. The issue is that it makes sense for many alternative nodes to include a centre node (if they include a A-node sufficiently close to the A=1 node), and it makes sense for CGN's to exclude a centre node because they never include a node that is close to A=1.

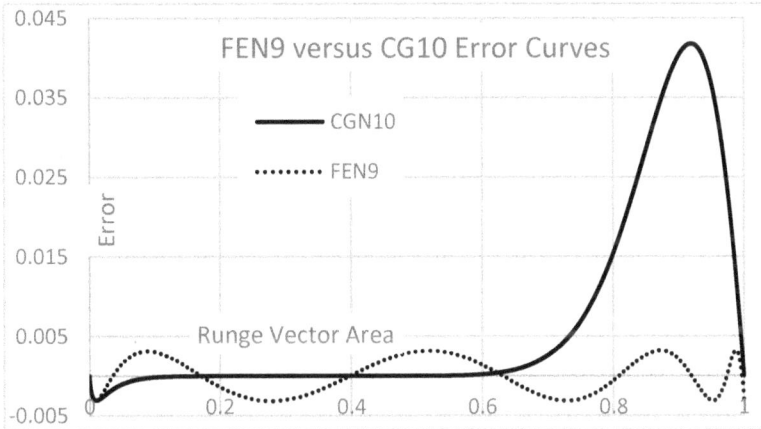

FEN9 versus CG10 Error Curves

Each set of FEN nodes comes with a maximum error limit for any Runge vector—on the chart above we can see that there are 9 Runge vectors (peaks and troughs) that achieve the maximum error limit. As noted above, the centre-node weight (~ 0.0031518 for the above FEN9 nodes) will equal the maximum error limit.

Suppose we constructed a curve from a linear combination of these worst-case vectors except that we exclude the b=∞ (A=1) bell-curve vector and the b=-1 extreme U-curve vector. The linear coefficients could be +1 for all the peaks and -1 for all the troughs so that the errors did not cancel out. This would construct a kind of average worst-case vector for the FEN nodes. The peaks and troughs occur at: b= {-0.991048089, -0.883951461, -0.318249654, 2.389104262, 16.0707585, 98.77805533, 797.5124428, 12471.34195} This means that the FEN9 integration error will be ~8*0.0031518 = 0.0252144. If we use CGN9 nodes we will get an error of ~-0.08024 and if we use CGN10 nodes we will get an error of ~ 0.00954.

Appendix B includes FEN17 and FEN33 x-nodes and weights.

Partial Spectrum FEN Nodes

These nodes control the maximum in-between error (Max IBE) to some pre-determined value (say 10^{-10}) adjacent to a nominated centre-point or centre B-node (say B=1/2 or 1/3). The size of the error-controlled region will depend on the nominated maximum error and the number of B-nodes.

Pick a centre node, and then expand the nodes out from that position towards A=0 and A=1 but depending on the number of nodes and the Max IBE we have selected, there may be substantial portions of the spectrum, at the edges adjacent A=0 and A=1, that are node free.

If we nominate a centre node rather than a centre point, then the centre node will not be at the exact centre because there is an even number of b-nodes. If we nominate a centre point, then there will be two nodes that are at the centre, one to the right and one to the left of the centre.

Appendix C includes a small selection of Partial Spectrum FEN x-nodes and weights.

Second Kind Nodes

If we consider an example of the Chebyshev second kind polynomial, say: $64*x^{\wedge 6}$ $-80*x^{\wedge 4}$ $+24*x^{\wedge 2}$ -1 and we integrate the areas between its adjacent roots ($x\sim\pm0.22252\ldots$, $x\sim\pm0.6234898\ldots$, $x\sim\pm0.9009688$) and between ±1 and the adjacent $\sim\pm0.9009688$ root then we will find all the areas are equal to $\sim\pm0.285714$. And in the same way that the full spectrum FEN nodes are like the 'Chebyshev nodes of the first kind', these Second Kind nodes are like 'Chebyshev nodes of the second kind' except that we are integrating the error with respect to A (or B instead of x) on the horizontal axis. So, for the sake of a name, I call them SKN's. The SKN17 x-Nodes and Weights are given below with some error chart comparisons with CGN18 nodes:

SKN17	
x-Nodes	Weights
6.33679561599284E-03	1.25103586184379E-02
3.04999868388404E-02	3.96720643503777E-02
9.64496337776384E-02	9.93013055968057E-02
2.45311229604688E-01	2.05699098958084E-01
5.07404109694010E-01	3.04571758593107E-01
7.96601757076238E-01	2.40340798806050E-01
9.55807712171684E-01	8.39734271935850E-02
9.96211531357274E-01	1.24609412618684E-02
0.00000000000000E+00	1.47024662168516E-03

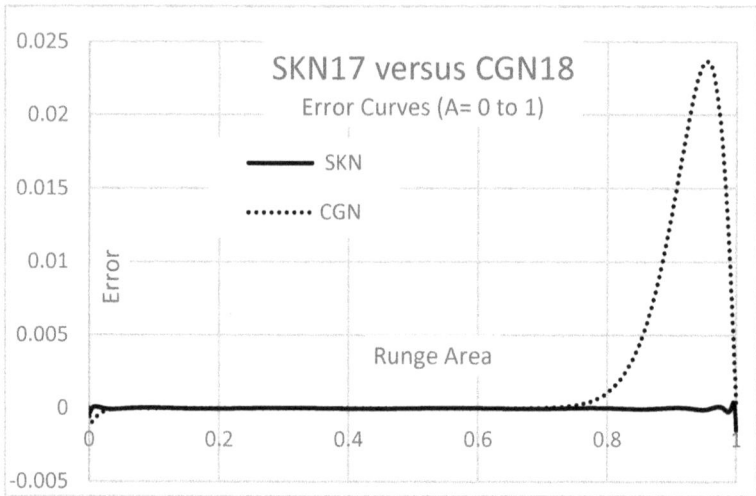

SKN17 versus CGN18

Error Curves (A= 0 to 1)

Runge Area

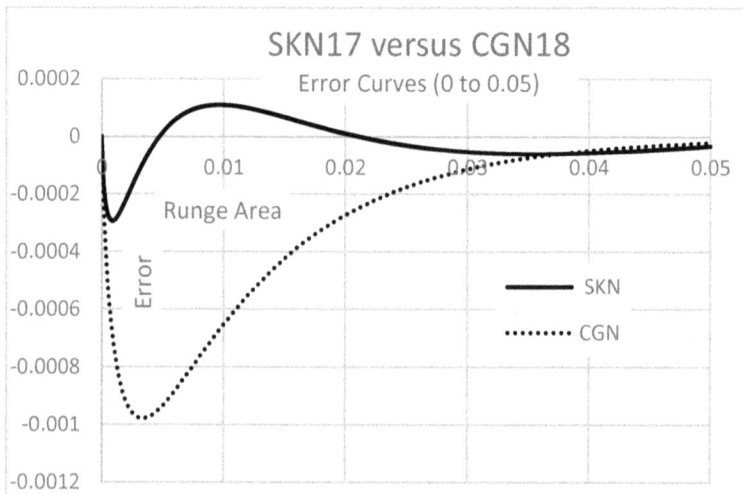

SKN17 versus CGN18

Error Curves (0 to 0.05)

Runge Area

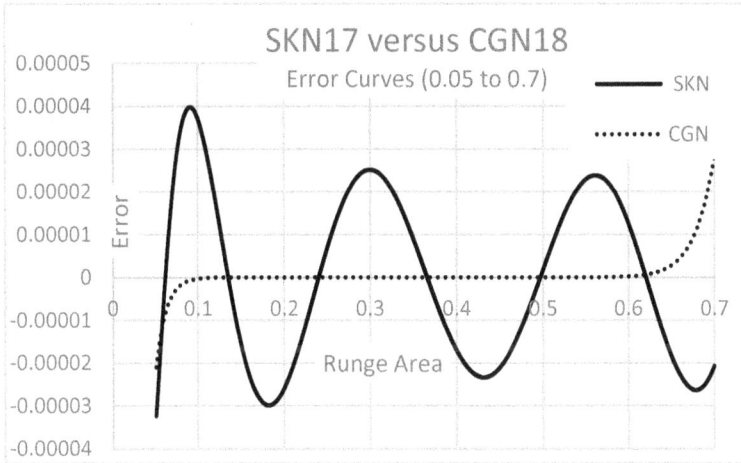

SKN17 versus CGN18
Error Curves (0.05 to 0.7)

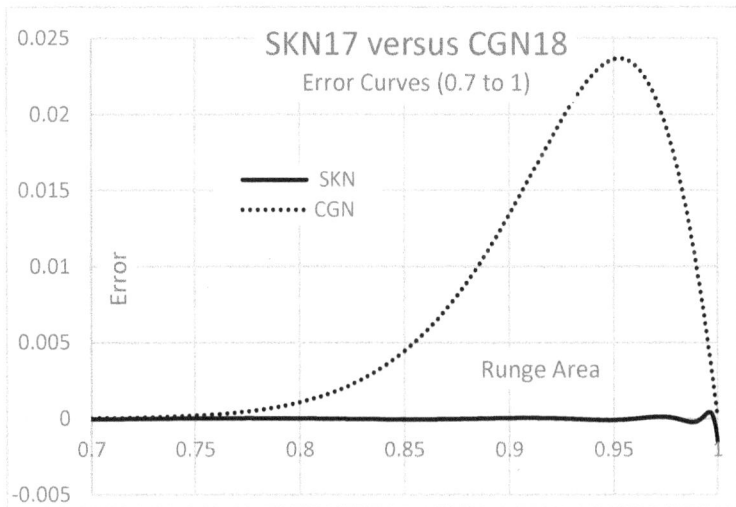

SKN17 versus CGN18
Error Curves (0.7 to 1)

The SKN17 nodes satisfy the condition $\int Abs(Error)\ dA$ between adjacent nodes is approximately equal to ±2.217E-05. Various approximations of the poly-normalised area A were used in the numerical integration process.

Minimum Total Error Nodes

'Total Error' means the 'integral of the absolute value of the error taken across the entire spectrum of Runge vectors with respect to vector poly-normalised area'. If we are using CGN's to do the integration then we will need to integrate the absolute value of the error between the nodes (with poly-normalised area or B-value as the vector shape parameter on the horizontal axis), like we did for the second kind nodes except we are now interested in minimising the sum of all errors rather than equating the errors.

MTN17	
xNodes	weights
8.43100218317372E-03	1.57244612202643E-02
3.66802335799436E-02	4.45145532422871E-02
1.07584696233001E-01	1.03732120564302E-01
2.58203568087213E-01	2.03346305410290E-01
5.12081043484581E-01	2.91396626306521E-01
7.89768298885789E-01	2.34352340984685E-01
9.49890940104783E-01	8.93266369146568E-02
9.95067499836519E-01	1.54935274149288E-02
0.00000000000000E+00	2.11342794206593E-03

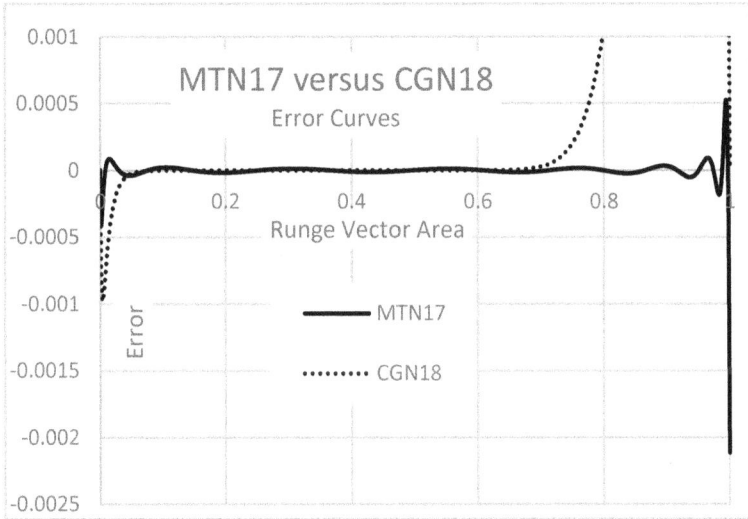

MTN17 versus CGN18
Error Curves

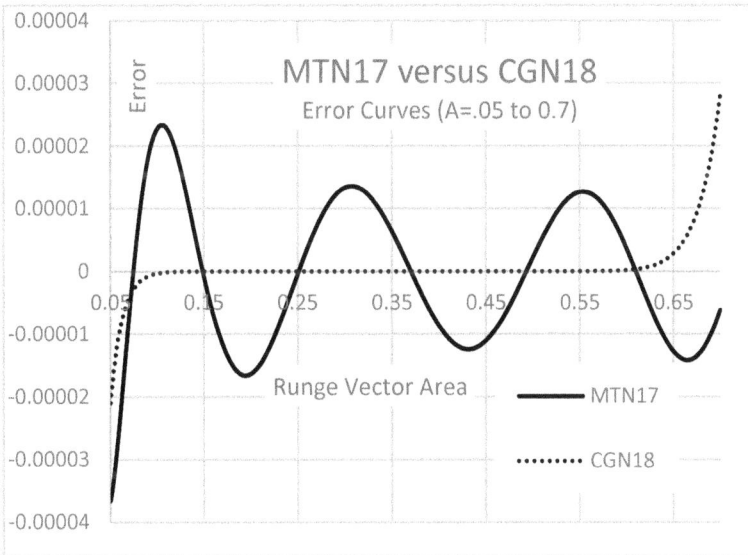

MTN17 versus CGN18
Error Curves (A=.05 to 0.7)

The error curve for these MTN nodes is somewhat like the error curve for the previous SKN nodes but with higher errors at the extremes and smaller errors for moderate vector shapes. The large spike at A=1 seems reasonable as these nodes will include a centre-node to integrate y=1 (the other A=1 vector) exactly. The CGN18 nodes generate an error of ~0.0236 at A~ 0.95 which is more than 10 times the magnitude of the MTN error and for a much larger portion of the total spectrum compared to the MTN spike near A=1.

For the high accuracy region of the CGN18 nodes—say from A= 0.142 to 0.55 then the CGN absolute error approaches ~$5*10^{-15}$ at A=1/3 and reaching a maximum of ~ $6*10^{-9}$ at the edges of the region whereas the MTN nodes show an error of up to about 0.00002.

8. CONCLUSION

The performance of our nodes is about how well the two-parameter vector shape parameters map the target curves. When we are integrating low order polynomials we want the two-parameter vector shape parameters (the polynomial index) to follow the pattern: 0,2,4,6,...

In some situations, it may be possible to curve fit a rational to a sample of the target curves to determine the range of vector shapes required. Or maybe it is possible to use Partial Spectrum FEN nodes to determine the range of vector shapes required.

In other situations, we may want to assess the performance of our nodes with an error curve. And although the Wikipedia comments about the performance of the Clenshaw Curtis nodes do not point to a Gauss node criterion, such comments suggest that in some situations the overall node performance may be more important than the CGN high-performance region.

Of course, there are other issues we must consider. Chebyshev nodes are very easy to generate. There is also the issue of complexity of a curve in the sense of the number of different vectors we need to accurately map a curve. The high-performance region of CGN's should mean that CGN's will be suitable for zooming in to small sections of a curve if we zoom in far enough.

From Section 5 (DE SOLUTIONS), the XTN nodes (based on a tangent curve fit to $1^{\wedge x}$), with the curious pattern of weights and x-nodes seem interesting. Most the solutions from the Section 7 (Iterative Solutions) seem interesting. Appendix D includes a discussion of some interesting Gauss Node Criterion.

APPENDICES

A: THE IR ASSESSMENT VECTOR

If we have CGN's that integrate 1, x^2, x^4, x^6, x^8,... we should expect that our RHS nodes will integrate the in-between shapes x^3, x^5, x^7 (from x=0 to x=1) reasonably well because such in-between vectors will be smooth curves similar in shape to the adjacent even powered vectors. However, this 'similarity of shape' thinking is not so good for low index vectors between x^1 and x^2 and much worse for vectors between x^0 and x^1.

We want to avoid the integer spacing of shape parameters associated with the polynomial vector which we can do with a power x^b vector, except that we may also want to avoid the non-smooth curve behaviour of small index power vectors. Gauss nodes designed to integrate power curves are Gauss nodes designed to integrate $(x^2)^{(b/2)}$ curves. The Runge vector avoids the power vector issues but getting to and from the Runge b-value and the Runge A-value has some inconveniences. We can get around these issues with the IR vector.

To get the IR vector we start by taking the first derivative of the odd rational $x^3/(1+a\,x^2)$ to get: $(x^2*(a*x^2 + 3))/(a*x^2 + 1)^2$. We now have a vector that integrates into a simple rational and guarantees a simple rational expression for the area of the vector. We can poly-normalise the vector by multiplying it by $(a+1)^2/(a+3)$ to get $(x^2 (a\,x^2 + 3))/(a\,x^2 + 1)^2 *(a+1)^2/(a+3)$ with an integral of $x^3/(1+a\,x^2)*(a+1)^2/(a+3)$. Now if we substitute a= (2/b - 1) then we will get the IR vector $(6\,b\,x^2 - 2\,(b - 2)\,x^4)/((b + 1)\,(b - (b - 2)\,x^2)^2)$ that has an area from x=0 to 1 equal to $1/(b+1)$. This final manipulation ensures the b-value mimics the power vector index. (Alternatively we could put $a = (1 - 3\,A)/(A - 1)$ we get a vector with poly-normalised area A as the shape parameter.)

For values of b equal to or larger than 2, the IR vector $(6 b x^2 - 2 (b - 2) x^4)/((b + 1) (b - (b - 2) x^2)^2)$ will generate U-curve vectors that look somewhat like polynomial vectors. For values of b smaller than 2 it will generate smooth inverted bell curves. But in all cases the area of the vector (from x=0 to 1) will be equal to $1/(b+1)$.

The following charts compare the IR assessment vector against the power vector assessment vector using CGN16 nodes. This is about comparing the IR vector with the power vector (using CGN's). Chart 1 of 3 is for b-values larger than 30 (A=0 to 1/31), Chart 2 of 3 is for b-values from 30 to 2 (A=1/31 to 1/3) and Chart 3 of 3 is for b values from 2 to 0 (A=1/3 to ∞):

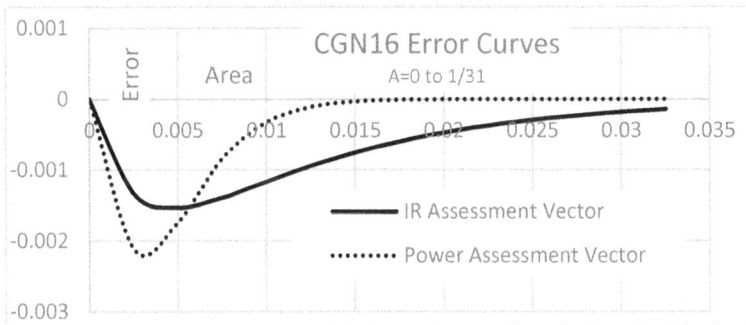

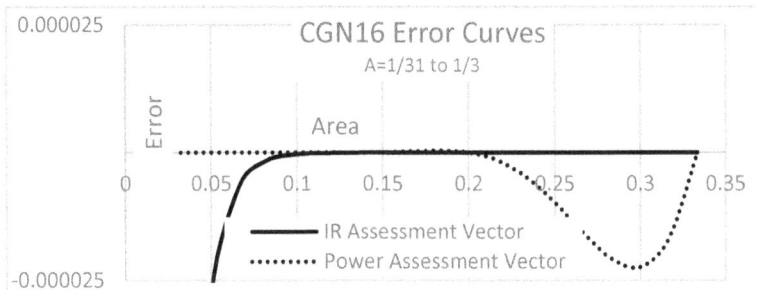

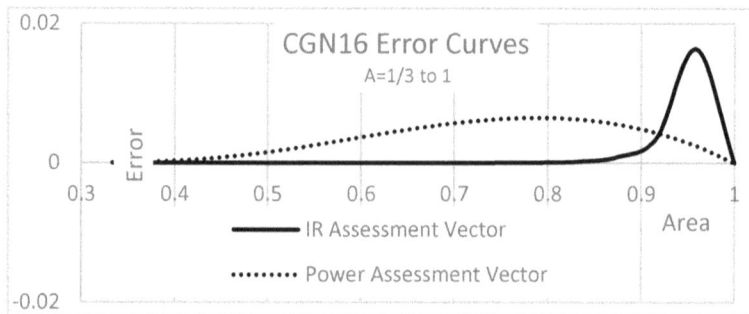

Because of the definition of error that we are using, both assessment vector curves show a pseudo-node at A=0. Except for a small region close to A=0, chart 1 shows a larger error for the IR assessment vector than for the power assessment vector.

According to the second chart, the IR assessment vector achieves a smaller error than the power vector between b=2 and b=4. If we count an error less than $5*10^{\wedge-15}$ as perfect performance (given the restraints of Excel calculation accuracy and 15 significant digit nodes) then these CGN16 nodes achieve perfect performance from A~ 0.25 to A~ 0.42 when assessed with the IR vector. However, against the power assessment vector there are no significant zones of perfect performance. Of course, the performance is perfect at the nodes (b =30, 28, 26, ...4, 2) but even in the middle of the best performance zone from b =24 to 20, when we plot the error against Vector Area on the horizontal axis, only a small fraction of the power vectors would be integrated with perfect performance.

The third chart shows a much better performance for the IR assessment vector for moderate bell-curve vectors and a much worse performance for more extreme bell-curve vectors.

B: GETTING CGN's WITH THE RUNGE VECTOR

Using the poly-normalised Runge vector we have:

$$\Sigma w_j \, x_j^2/(1+b \, x_j^2) \, *(1+b) \sim (1+b) \, *(1/b - atan(\sqrt{b})/b^{(3/2)})_{\{for \, j \, =1,2,3...\}}$$

If we cancel the (1+b) factor on both sides of the above expression, then the TEX of the target curve on the RHS at b=0 will be:

$$= 1/3 - b/5 + b^2/7 - b^3/9 + b^4/11 - ...$$

This means we want:

$$TEX\{\Sigma w_j \, x_j^2/(1+b \, x_j^2)\} \sim 1/3 - b/5 + b^2/7 - b^3/9 + b^4/11 - ..._{\{at \, b=0\}}$$

And because $x^2/(1+b \, x^2)$ can be written as the series: $x^2 - b \, x^4 + b^2 \, x^6 - b^3 \, x^8 + ...$ then:

$$w_1 \, *x_1^2/(1+ b* \, x_1^2) + w_2 \, *x_2^2/(1+ b \, *x_2^2) + w_3 \, *x_3^2/(1+ b \, *x_3^2) + ...$$ can be written as:

$$(w_1 \, x_1^2 + w_2 \, x_2^2 + w_3 \, x_3^2 + ...) + b \, (-w_1 \, x_1^4 - w_2 \, x_2^4 - w_3 \, x_3^4 + ...) + b^2$$
$$(w_1 \, x_1^6 + w_2 \, x_2^6 + w_3 \, x_3^6 + ...) + b^3 \, (-w_1 \, x_1^8 - w_2 \, x_2^8 - w_3 \, x_3^8 + ...) +$$
$$b^4 \, (w_1 \, x_1^{10} + w_2 \, x_2^{10} + w_3 \, x_3^{10} + ...) + b^5 \, (-w_1 \, x_1^{12} - w_2 \, x_2^{12} - w_3 \, x_3^{12} + ...) + ...$$

We now need to solve the simultaneous equations:

b^0: $\quad w_1 \, x_1^2 + w_2 \, x_2^2 + w_3 \, x_3^2 + ... = 1/3$, $\qquad b^1$: $\quad -w_1 \, x_1^4 - w_2 \, x_2^4 - w_3 \, x_3^4 + ... = -1/5$,

b^2: $\quad w_1 \, x_1^6 + w_2 \, x_2^6 + w_3 \, x_3^6 + ... = 1/7$, $\qquad$ etc

These sorts of '$\Sigma w_j \, *x_j^b$ = constant' simultaneous equations are the sorts of equations that difference equation theory solves. In this case we already have the solution because these equations are the CGN DE equations. For example, the RHS value for the expression in x^n is equal $1/(1+n)$ which is the integral of x^n from x=0 to 1. Hence CGN nodes create a tangent to the Runge area curve (or rather, the error curve will be tangent to the horizontal axis when we assess CGN's with the Runge vector) at b= 0 (A=1/3).

We could use CGN nodes and weights in a similar way that we use a Taylor expansion to determine values of A(b)/(1+b) for values of b close to zero. Convergence may be faster, but the accuracy of the calculation will depend on the accuracy of the Gauss parameters whereas the TEX parameters are exact.

C: FLAT ERROR (FEN) NODES

Full-spectrum FEN Nodes

Full-spectrum FEN nodes are almost independent of arbitrary criteria. The maximum in-between absolute error value (Max IBE) is not shown for the full spectrum FEN nodes because it is equal to the weight associated with the centre node.

FEN17	
x-Nodes	Weights
0.00000000000000E+00	1.72008901045251E-04
1.42393377181518E-03	3.57158039251072E-03
1.04712095270273E-02	1.73737234909627E-02
4.57249322435639E-02	6.05211742306655E-02
1.53172455733828E-01	1.68280777503636E-01
4.03975413185646E-01	3.30708053622080E-01
7.51355298997290E-01	3.10033080214145E-01
9.54263667542138E-01	9.90909663794750E-02
9.97207609340765E-01	1.02486352654779E-02

FEN33	
x-Nodes	Weights↓
0.00000000000000E+00	2.63368993969512E-06
2.18025015737131E-05	5.46866871506960E-05
1.60359661194599E-04	2.66157083337696E-04
7.01906456761372E-04	9.33218710265604E-04
2.39651502854852E-03	2.73396362948286E-03
7.01536734812567E-03	7.11852133765672E-03
1.84406719087575E-02	1.70100429149113E-02
4.46717438827489E-02	3.79445865612622E-02
1.01070996963676E-01	7.91881690450269E-02
2.13198923932343E-01	1.50224314626688E-01

4.06952956647119E-01	2.34714529275188E-01
6.59319240100308E-01	2.49815508197297E-01
8.66407628875343E-01	1.52347498621144E-01
9.64726641936651E-01	5.33974463510876E-02
9.93488586127477E-01	1.21936383387172E-02
9.99206869337849E-01	1.89434670562762E-03
9.99956648014613E-01	1.60738225214712E-04

Partial-spectrum FEN Nodes

Partial spectrum FEN solutions depend on the nomination of some arbitrary centre position and on the nomination of some arbitrary maximum in-between absolute error value (Max IBE) for the nodes. The larger the Max IBE the larger the spread of the flat-error zone for the nodes.

FEN17, Partial Spectrum with Max IBE 10^-7 Centre Node @ b=0 (B=1/3)	
x-Nodes	Weights
9.96571151755592E-01	9.39804728245207E-03
9.76483924733685E-01	3.45144999848932E-02
9.17155569244353E-01	9.01244614066254E-02
7.87555489661680E-01	1.69479362874113E-01
5.89522406570837E-01	2.15647293516163E-01
3.80067840974702E-01	1.93648363362177E-01
2.12137245110582E-01	1.41644465595242E-01
9.23978645726744E-02	1.01668538840183E-01
0.00000000000000E+00	4.38749671381485E-02

FEN17, Partial Spectrum with Max IBE $10^{\wedge-10}$ Centred @ b=0 (B=1/3)	
x-Nodes	Weights
9.92974932638042E-01	1.83944477743904E-02
9.59916750738778E-01	4.95329573538775E-02
8.89932701844043E-01	9.21609457198528E-02
7.74285077533280E-01	1.38234935124950E-01
6.18849550857079E-01	1.68447384755573E-01
4.46727960103716E-01	1.71458873564586E-01
2.82136028059794E-01	1.56115122872643E-01
1.34888622785409E-01	1.39355092783991E-01
0.00000000000000E+00	6.63002400501337E-02

FEN33, Partial Spectrum with Max IBE $10^{\wedge -10}$ Centred @ b=2 (B=1/2)	
x-Nodes	Weights
9.99333040088518E-01	1.77830486148299E-03
9.95874015593077E-01	5.57253000594828E-03
9.86820274310550E-01	1.35235649392462E-02
9.66086387895698E-01	2.97242241183817E-02
9.23120925537483E-01	5.85160575324407E-02
8.45051126070776E-01	9.88371592906074E-02
7.26058633912655E-01	1.37022043690310E-01
5.78545444928383E-01	1.53150668648530E-01
4.29338747254124E-01	1.41229356412391E-01
3.01517054846411E-01	1.13047755499982E-01
2.03838427823737E-01	8.28015487267697E-02
1.34117160941895E-01	5.77513978376468E-02
8.60715472627736E-02	3.94223823134505E-02
5.33094822019913E-02	2.69721492524750E-02
3.05998860386202E-02	1.91142242083862E-02
1.39002177994751E-02	1.48130524965133E-02
0.00000000000000E+00	6.72358016543817E-03

FEN33, Partial Spectrum with Max IBE $10^{\wedge-15}$ Centred @ b=2 (B=1/2)	
x-Nodes	Weights↓
9.969848166011600E-01	7.813883929130030E-03
9.834580539804380E-01	1.965640467174030E-02
9.565996428926410E-01	3.468437541953060E-02
9.128237676136640E-01	5.343666604796670E-02
8.489740104861750E-01	7.435524058037310E-02
7.647174030808390E-01	9.342286553207660E-02
6.644192018906520E-01	1.056899452554340E-01
5.566355174599820E-01	1.081990876374560E-01
4.510641533313290E-01	1.016713066109150E-01
3.551918312773450E-01	8.945445020416450E-02
2.727980572995350E-01	7.526859833035690E-02
2.044077718376680E-01	6.177890266542820E-02
1.485325707772800E-01	5.038262745274280E-02
1.027706681186210E-01	4.158726974654930E-02
6.447302096904790E-02	3.544206848386530E-02
3.104108846657930E-02	3.183346042137020E-02
0.000000000000000E+00	1.532284701089820E-02

D: ITERATING X-NODES AND WEIGHTS

Once we have a set of valid Gauss nodes (nodes with the appropriate number of intersections on the vector area curve) it is possible to increment the x-node values and the weights with small increments to get other Gauss node solutions.

Iteration of the x-nodes and weights as the independent variables was used to find some of the FEN nodes. By taking the first derivative of the error equation (expressed as the sum of the partial fraction vectors minus A(b)) with respect to the b-variable means that we then need only to find the zeros to get the position of the maximums (peaks) and minimums (troughs).

Equal Weight Gauss Nodes

Suppose we have 'n' vector shape parameters then we will have only 'n+1' free variables (as the weights are predetermined to be equal) and we want to find '2n' curve fit points, which seems like an impossible problem. However, this is not necessarily an impossible because we can use the Runge vector and B-nodes to first find an equal weight solution.

Suppose we want to find N4 equal weight nodes using a Runge rational. We can start with 3 free B-nodes and one B node fixed at B=1; we can then shift the free nodes around as we choose because every set of B-nodes will generate a valid Gauss node solution. So, we shift the B-nodes around until the weights are both equal to 0.5—we have only 2 weight conditions (w_1 =0.5, w_2 =0.5) that we want to achieve, and we have 3 free B-nodes to achieve these conditions. Because the weights must add to 1 there is only one weight condition that we need to satisfy. This works because by insisting on the B=1 node we have already enforced the condition that the weights must add to 1. The following is one of many possible equal weight Gauss node solutions:

x-Nodes	Weights
0.1993	0.5
0.8079	0.5

If we plot the error curve for these nodes, then we will see that the spacing the B-nodes is B={ 0.19982642, 0.462524653, 0.783698885, 1.0}. So not only is a solution possible but it seems that we have a stable solution that will tolerate moderate changes in the x-Node values without the loss of an intersection.

Suppose we now want to make the 0.19982642 B-node equal to 0.2 then we need to be able to change the x-nodes because if we try and modify the B-nodes we will immediately destroy the equal weight condition. This means can only modify x_1 or x_2 or both. (One solution to this problem is: x_1= 0.199295014981398, x_2 = 0.807864317159906 which will give the B-nodes: 0.2, 0.462401489, 0.783719059, 1.0.)

Or suppose we also want to find some new x-nodes of our choosing. And we can in fact make the x-nodes equal to 0.2 and 0.8 to get:

x-Nodes	Weights
0.2	0.5
0.8	0.5

Double and Triple Gauss Nodes
The above nodes are valid Gauss nodes not just when assessed with the Runge vector but also when assessed with the IR vector and the power vector. However, as we can see from the tangent

at A=0.5 on the power vector error curve below, we are close to losing two power curve nodes.

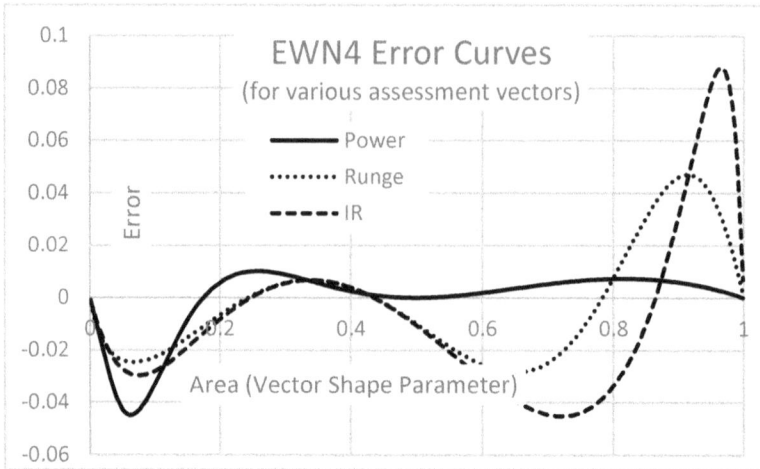

The A=0.5 power vector node means that the nodes will integrate $y = (x^2)^{(1/2)}$ non-smooth curve exactly. We could adjust the x-node values so that the tangent became 2 distinct intersection points one with one intersection through or to the left of $A = 1/3$ say, but then we would lose those nice Newton-Cotes like x-node values 0.2 and 0.8.

Progressive Nodes

Suppose we wanted to construct progressive Gauss nodes based on initial CGN5 nodes. This means we want to reuse the CGN5 x-node values (namely $x_0 = 0$, $x_1 = 0.5384693101056831$, $x_2 = 0.906179845938664$) in a higher order set of Gauss nodes. (The $x_0 = 0$ node part of the problem is trivial because we can ignore the centre node until the end of the process.)

In the interest of the progressive nodes having sufficient free parameters, we choose our progressive nodes to include an x-node between the CGN5 nodes and one to the left and one to the right of the CGN5 nodes and identify them as PRN11 nodes.

We now write out ten B-node values spaced out somewhat uniformly from B= 0 to 1. We need to iterate these B-node values until we get x-nodes that are identical to the CGN5 x-nodes. And once we have achieved the CGN5 x-nodes with our new Runge11 nodes, we will need to preserve the values if we want to make any further changes. This means we will only be able to iterate the non-preserved values which includes all weights and the non-CGN5 x-nodes. For each increment we will need to watch the error curve to ensure that we do not lose any intersections. So again, at the final stages of this problem we need to switch from iterating B-nodes to iterating x-nodes because of the need to preserve two x-node values.

Although polynomial vectors are extremely restrictive, it was possible to increment the non-preserved values around until the corresponding x-nodes and weights generated intersections on the power area curve at x^2, x^4, x^6, x^8, x^{12}, x^{18}, x^{28}, x^{48}. There are only 8 intersections (instead of the 10 required for power curve Gauss Nodes) and the last 4 intersections were selected according to whatever seemed achievable. But they are Gauss nodes because they have the required number of intersections on the Runge area curve: {B~0.152762, 0.188640, 0.225195, 0.262414, 0.304264, 0.321131, 0.363854, 0.401020, 0.437402, 0.472909} which is to say they are valid Gauss nodes when assessed with the Runge vector. The x-nodes in bold are the CGN5 x-nodes:

PRN11 x-nodes	weights
0.0000000000000000	0.138936404946906

0.276457894820851	0.273008897957268
0.5384693101056831	0.245688751579280
0.756815143799123	0.186419855142624
0.906179845938664	0.111915868301082
0.983089833842477	0.0440302220728386

(Not only are there only 8 polynomial intersections but the x^2 intersection is closer to $x^{2.000000000044}$ which means we get an x^2 integration error of -2.89E-14.)

High Order Solutions
The above successful attempts to find equal weight nodes and progressive nodes do not mean that higher order solutions will be possible. And if higher order solutions are possible, they might prove to be slow and difficult problems.

Finding the solution to the above two examples with only 2 equal weights and the second with only 2 preserved x-nodes was a very slow process. Once a solution, was found that it was easy to play with the free x-nodes and free weights to find alternative Runge vector solutions. We can do that with CGN nodes, providing we keep an eye on the intersections with the Area curve and of course the process is going to much easier if we understand our vectors to be power vectors rather than polynomial vectors.

www.ingramcontent.com/pod-product-compliance
Lightning Source LLC
Chambersburg PA
CBHW040929210326
41597CB00030B/5236